CURRENT STRATEGIES FOR ENGINEERING CONTROLS IN
Nanomaterial Production and Downstream Handling Processes

DEPARTMENT OF HEALTH AND HUMAN SERVICES
Centers for Disease Control and Prevention
National Institute for Occupational Safety and Health

This document is in the public domain and may be freely copied or reprinted.

Disclaimer

Mention of any company or product does not constitute endorsement by the National Institute for Occupational Safety and Health (NIOSH). In addition, citations to Web sites external to NIOSH do not constitute NIOSH endorsement of the sponsoring organizations or their programs or products. Furthermore, NIOSH is not responsible for the content of these Web sites. All Web addresses referenced in this document were accessible as of the publication date.

Ordering information

To receive documents or more information about occupational safety and health topics, contact NIOSH:

Telephone: **1–800–CDC–INFO** (1-800-232-4636)
TTY: 1–888–232–6348
CDC INFO: www.cdc.gov/info

or visit the NIOSH web site at www.cdc.gov/niosh

For a monthly update on news at NIOSH, subscribe to *NIOSH eNews* by visiting www.cdc.gov/niosh/eNews.

Suggested Citation

NIOSH [2013]. Current strategies for engineering controls in nanomaterial production and downstream handling processes. Cincinnati, OH: U.S. Department of Health and Human Services, Centers for Disease Control and Prevention, National Institute for Occupational Safety and Health, DHHS (NIOSH) Publication No. 2014–102.

DHHS (NIOSH) Publication No. 2014–102

November 2013

SAFER • HEALTHIER • PEOPLE™

Foreword

The National Institute for Occupational Safety and Health (NIOSH) is charged with protecting the safety and health of workers through research and training. An area of current concentration is the study of nanotechnology, the science of matter near the atomic scale. Much of the current research focuses on understanding the toxicology of emerging nanomaterials as well as exposure assessment; very little research has been conducted on hazard control for exposures to nanomaterials. As we continue to research the health effects produced by nanomaterials, particularly as new materials and products continue to be introduced, it is prudent to protect workers now from potential adverse health outcomes. Controlling exposures to occupational hazards is the fundamental method of protecting workers. Traditionally, a hierarchy of controls has been used as a means of determining how to implement feasible and effective control solutions.

- Elimination
- Substitution
- Engineering Controls
- Administrative Controls
- Personal Protective Equipment

Following this hierarchy normally leads to the implementation of inherently safer systems, where the risk of illness or injury has been substantially reduced. Engineering controls are favored over administrative and personal protective equipment for controlling existing worker exposures in the workplace because they are designed to remove the hazard at the source, before it comes in contact with the worker. However, evidence of control effectiveness for nanomaterial production and downstream use is scarce. This document is a summary of available technologies that can be used in the nanotechnology industry. While some of these have been evaluated in this industry, others have been shown to be effective at controlling similar processes in other industries. The identification and adoption of control technologies that have been shown effective in other industries is an important first step in reducing worker exposures to engineered nanoparticles.

Our hope is that this document will aid in the selection of engineering controls for the fabrication and use of products in the nanotechnology field. As this field continues to expand, it is paramount that the health and safety of workers is protected.

John Howard, M.D.
Director, National Institute for
   Occupational Safety and Health
Centers for Disease Control and Prevention

Executive Summary

The focus of this document is to identify and describe strategies for the engineering control of worker exposure during the production or use of engineered nanomaterials. Engineered nanomaterials are materials that are intentionally produced and have at least one primary dimension less than 100 nanometers (nm). Nanomaterials may have properties different from those of larger particles of the same material, making them unique and desirable for specific product applications. The consumer products market currently has more than 1,000 nanomaterial-containing products including makeup, sunscreen, food storage products, appliances, clothing, electronics, computers, sporting goods, and coatings. As more nanomaterials are introduced into the workplace and nano-enabled products enter the market, it is essential that producers and users of engineered nanomaterials ensure a safe and healthy work environment.

The toxicity of nanoparticles may be affected by different physicochemical properties, including size, shape, chemistry, surface properties, agglomeration, biopersistence, solubility, and charge, as well as effects from attached functional groups and crystalline structure. The greater surface-area-to-mass ratio of nanoparticles makes them generally more reactive than their macro-sized counterparts. These properties are the same ones that make nanomaterials unique and valuable in manufacturing many products. Though human health effects from exposure have not been reported, a number of laboratory animal studies have been conducted. Pulmonary inflammation has been observed in animals exposed to nano-sized TiO_2 and carbon nanotubes (CNTs). Other studies have shown that nanoparticles can translocate to the circulatory system and to the brain causing oxidative stress. Of concern is the finding that certain types of CNTs have shown toxicological response similar to asbestos in mice. These animal study results are examples, and further toxicological studies need to be conducted to establish the potential health effects to humans from acute and chronic exposure to nanomaterials.

Currently, there are no established regulatory occupational exposure limits (OELs) for nanomaterials in the United States; however, other countries have established standards for some nanomaterials, and some companies have supplied OELs for their products. In 2011, NIOSH issued a recommended exposure limit (REL) for ultrafine (nano) titanium dioxide and a draft REL for carbon nanotubes and carbon nanofibers. Because of the lack of regulatory standards and formal recommendations for many nanomaterials in the United States, it is difficult to determine or even estimate a safe exposure level.

Many of the basic methods of producing nanomaterials occur in an enclosure or reactor, which may be operated under positive pressure. Exposure can occur due to leakage from the reactor or when a worker's activities involve direct manipulation of nanomaterials. Batch-type processes involved in the production of nanomaterials include operating reactors, mixing, drying, and thermal treatment. Exposure-causing activities at production plants and laboratories employing nanomaterials include harvesting (e.g., scraping materials out of reactors), bagging, packaging, and reactor cleaning. Downstream activities that may release nanomaterials include bag dumping, manual transfer between processes, mixing or compounding, powder sifting, and machining of parts that contain nanomaterials.

Hazards involved in manufacturing and processing nanomaterials should be managed as part of a comprehensive occupational safety, health, and environmental management plan. Preliminary hazard assessments (PHAs) are frequently conducted as initial risk assessments to determine whether more sophisticated analytical methods are needed. PHAs are important so that the need for control measures is realized, and the means for risk mitigation can be designed to be part of the operation during the planning stage.

Engineering controls protect workers by removing hazardous conditions or placing a barrier between the worker and the hazard, and, with good safe handling techniques, they are likely to be the most effective control strategy for nanomaterials. The identification and adoption of control technologies that have been shown effective in other industries are important first steps in reducing worker exposures to engineered nanoparticles. Properly designing, using, and evaluating the effectiveness of these controls is a key component in a comprehensive health and safety program. Potential exposure control approaches for commonly used processes include commercial technologies, such as a laboratory fume hood, or techniques adopted from the pharmaceutical industry, such as continuous liner product bagging systems.

The assessment of control effectiveness is essential for verifying that the exposure goals of the facility have been successfully met. Essential control evaluation tools include time-tested techniques, such as airflow visualization and measurement, as well as quantitative containment test methods, including tracer gas testing. Further methods, such as video exposure monitoring, provide information on critical task-based exposures, which will help to identify high-exposure activities and help provide the basis for interventions.

This page left intentionally blank

Contents

Foreword	iii
Executive Summary	iv
List of Figures	x
List of Tables	xi
List of Abbreviations	xii
Acknowledgements	xiv
1 Introduction	**1**
1.1 Background	1
1.2 Industry Overview	3
1.3 Occupational Safety and Health Management Systems	3
1.3.1 Prevention through Design (PtD)	6
1.3.2 OELs as Applied to Nanotechnology	7
1.3.3 Control Banding	7
2 Exposure Control Strategies and the Hierarchy of Controls	**9**
2.1 Elimination	9
2.2 Substitution	10
2.3 Engineering Controls	10
2.3.1 Ventilation	11
2.3.1.1 Local Exhaust Ventilation	12
2.3.1.2 Air Filtration	14
2.3.2 Nonventilation Engineering Controls	16
2.4 Administrative Controls	17
2.5 Personal Protective Equipment (PPE)	18
2.5.1 Skin Protection	18
2.5.2 Respiratory Protection	18
3 Nanotechnology Processes and Engineering Controls	**21**
3.1 Primary Nanotechnology Production and Downstream Processes	21
3.2 Engineering Control Approaches to Reducing Exposures	22
3.3 Ventilation and General Considerations	24
3.4 Exposure Control Technologies for Common Processes	25
3.4.1 Reactor Operation and Cleanout Processes	27

  3.4.2 Small-scale Weighing and Handling of Nanopowders 30
    3.4.2.1 Fume Hood Enclosures 31
    3.4.2.2 Biological Safety Cabinets 33
    3.4.2.3 Glove Box Isolators 34
    3.4.2.4 Air Curtain Fume Hood 35
    3.4.2.5 Summary 36
  3.4.3 Intermediate and Finishing Processes 37
    3.4.3.1 Product Discharge/Bag Filling. 38
    3.4.3.2 Bag Dumping/Emptying 41
    3.4.3.3 Large-scale Material Handling/Packaging. 43
    3.4.3.4 Nanocomposite Machining 44
    3.4.3.5 Summary 44
  3.4.4 Maintenance Tasks 45
    3.4.4.1 Filter Change-out—Bag In/Bag Out Systems. 46
    3.4.4.2 Spill Cleanup Procedures 46

4 Control Evaluations. 47
 4.1 Approaches to Evaluation 47
 4.1.1 Identification of Emission Sources 47
 4.1.2 Background and Area Monitoring 47
 4.1.3 Air Monitoring and Filter Sampling 48
 4.1.4 Assessment of Air Velocities and Patterns 50
 4.1.5 Facility Sampling and Evaluation Checklist 52

4.2 Evaluating Sources of Emissions and Exposures to Nanomaterials. 57
 4.2.1 Direct-reading Monitoring 57

    4.2.2 Off-line Analysis 58
    4.2.3 Video Exposure Monitoring..................................... 58

4.3 Evaluating Ventilation Control Systems 59
    4.3.1 Standard Containment Test Methods for Ventilated Enclosures 59

5 Conclusions and Recommendations 61
    5.1 General 61
    5.2 Control Banding 61
    5.3 Hierarchy of Controls 62
    5.4 Engineering Controls 62
    5.5 Administrative Controls 62
    5.6 Personal Protective Equipment 63

References 65

Appendix A: Sources for Risk Management Guidance 75

Appendix B: Sources of Guidance for Control Design 77

List of Figures

Figure 1. Atomic structure of a spherical fullerene

Figure 2. How control measures are incorporated into an occupational safety and health management system

Figure 3. Worker reaching into drum

Figure 4. Graphical representation of the hierarchy of controls

Figure 5. Four primary filter collection mechanisms

Figure 6. Collection efficiency curve: fractional collection efficiency versus particle diameter for a typical filter

Figure 7. A large-scale ventilated reactor enclosure used to contain production furnaces to mitigate particle emissions in the workplace

Figure 8. A canopy hood used to control emissions from hot processes

Figure 9. Schematic illustration of how wakes caused by the human body can transport air contaminants into the worker's breathing zone

Figure 10. Nano containment hood adapted from a pharmaceutical balance enclosure

Figure 11. A tabletop model of a Class II, Type A2 biological safety cabinet (BSC)

Figure 12. A glove box isolator for handling substances that require a high level of containment

Figure 13. Air curtain safety cabinet hood that uses push-pull ventilation

Figure 14. Ventilated collar-type exhaust hoods for containing dust during product discharge or manual bag filling

Figure 15. An inflatable seal is used to contain nanopowders/dusts as they are discharged from a process such as spray drying

Figure 16. A continuous liner product off-loading system that uses a continuous feed of bag liners fitted to the process outlet to isolate and contain process emissions and product

Figure 17. A ventilated bag-dumping station that reduces dust emissions during the emptying of product from bags into a process hopper

Figure 18. A laminar downflow booth for handling large quantities of powders

Figure 19. Bag in/bag out procedures. This photo shows the removal of a dirty air filter from a ventilation unit into a plastic bag to minimize worker exposure to particles captured by the filter unit

Figure 20. Operating principle of a Pitot tube (left) and different types of Pitot tubes (right)

Figure 21. Smoke generator to visualize airflow

List of Tables

Table 1. Potential sources of emission from production and downstream processes

Table 2. Process/tasks and emission

Table 3. Summary of instruments and techniques for monitoring nanoparticle emissions in nanomanufacturing workplaces

Table 4. Checklist of controls for nanomaterial manufacturing and handling

Table 5. Comparison of the fume hood performance test methods

List of Abbreviations

ACGIH	American Conference of Governmental Industrial Hygienists
AIHA	American Industrial Hygiene Association
ANSI	American National Standards Institute
APF	assigned protection factor
ASHRAE	American Society of Heating, Refrigerating, and Air Conditioning Engineers
BSC	biological safety cabinet
BSI	British Standards Institute
CAV	constant air volume
CDC	Centers for Disease Control and Prevention
cfm	cubic feet per minute
CNF	carbon nanofiber
CNT	carbon nanotube
CPC	condensation particle counter
CVD	chemical vapor deposition
DMPS	differential mobility particle sizer
ELPI	electrical low pressure impactor
EPA	Environmental Protection Agency
FFR	filtering facepiece respirator
FMPS	fast mobility particle sizer
fpm	feet per minute
HEPA	high efficiency particulate air
HSE	Health and Safety Executive
IH	industrial hygiene
kg	kilogram
lbs	pounds
LEV	local exhaust ventilation
LPM	liters per minute
MPPS	most penetrating particle size
MSDS	material safety data sheet
MUC	maximum use concentration
NIOSH	National Institute for Occupational Safety and Health
nm	nanometer
OEL	occupational exposure limit
PEL	permissible exposure limit

PHA	preliminary hazard assessment
PM	preventive maintenance
PPE	personal protective equipment
PtD	prevention through design
R&D	research and development
REL	recommended exposure limit
SMACNA	Sheet Metal and Air Conditioning Contractors' National Association
SMPS	scanning mobility particle sizer
SOP	standard operating procedures
TEM	transmission electron microscopy
TEOM	tapered element oscillating microbalance
TLV®	threshold limit value
TWA	time- weighted average
VAV	variable air volume
VEM	video exposure monitoring
wg	water gauge
µg	microgram
µm	micrometer

Acknowledgments

This document was developed by the NIOSH Division of Applied Research and Technology (DART), Gregory Lotz, PhD, Director. Jennifer L. Topmiller, MS, was the project officer for this document, assisted in great part by Kevin H. Dunn, ScD, CIH. Other members of DART instrumental in the production of this document include Scott Earnest, PhD, PE, CSP; Liming Lo, PhD; Ron Hall, MS, CIH, CSP; Mike Gressel, PhD, CSP; Alan Echt, DrPh, CIH; and William Heitbrink, PhD, CIH (contractor). Elizabeth Fryer also provided writing and editing support in the initial stages.

The authors gratefully acknowledge the contributions of the following NIOSH personnel who assisted with the technical content and review of the document.

Division of Respiratory Disease Studies
Stephen B. Martin, Jr., MS, PE

Education and Information Division
Charles Geraci, PhD, CIH

Laura Hodson, MSPH, CIH

Health Effects Laboratory Division
Bean T. Chen, PhD

National Personal Protective Technology Laboratory
Pengfei Gao, PhD, CIH

Office of the Director
Paul Middendorf, PhD, CIH

The authors also wish to thank Cathy Rotunda, EdD, Brenda J. Jones, and Vanessa Williams for their assistance with editing and layout for this report. Cover photographs are courtesy of Quantum Sphere, Inc. and Bon-ki Ku, PhD, of NIOSH.

Special appreciation is expressed to the following who served as independent, external reviewers. Their input contributed greatly to the improvement of this document.

Keith Swain, DuPont, Wilmington, Delaware

Richard Prodans, CIH, CSP, Abbott, Abbott Park, Illinois

John Weaver, Birck Nanotechnology Center, Purdue University, West Lafayette, Indiana

Gurumurthy Ramachandran, PhD, CIH, University of Minnesota, Minneapolis, Minnesota

Phil Demokritou, PhD, Harvard University, Boston, Massachusetts

CHAPTER 1
Introduction

The number of commercial applications of nanomaterials is growing at a tremendous rate. As this rapid growth continues, it is essential that producers and users of nanomaterials ensure a safe and healthy work environment for employees who may be exposed to these materials. Unfortunately, because nanotechnology is so new, we do not know or fully understand how occupational exposures to these agents may affect the health and safety of workers or even what levels of exposure may be acceptable. Given our current knowledge in this field, it is important to take precautions to minimize exposures and protect safety and health.

This document discusses approaches and strategies to protect workers from potentially harmful exposures during nanomaterial manufacturing, use, and handling processes. Its purpose is to provide the best available current knowledge of how workers may be exposed and provide guidance on exposure control and evaluation. It is intended to be used as a reference by plant managers and owners who are responsible for making decisions regarding capital allocations, as well as health and safety professionals, engineers, and industrial hygienists who are specifically charged with protecting worker health in this new and growing field. Because little has been published on exposure controls in the production and use of nanomaterials, this document focuses on applications that have relevance to the field of nanotechnology and on engineering control technologies currently used, and known to be effective, in other industries. This document also addresses other approaches to worker protection, such as the use of administrative controls and personal protective equipment.

1.1 Background

Nanotechnology is the manipulation of matter at the atomic scale to create materials, devices, or systems with new properties and/or functions. Around the world, the introduction of nanotechnology promises great societal benefits across many economic sectors: energy, healthcare, industry, communications, agriculture, consumer products, and others [Sellers et al. 2009].

Some nanoparticles are natural, as in sea salt or pine tree pollen, or are incidentally produced, as in volcanic explosions or diesel engine emissions. The focus of this document is engineered nanomaterials, those materials deliberately engineered and manufactured to have certain properties and have at least one primary dimension of less than 100 nanometers (nm). Nanomaterials have properties different from those of their bulk components. For example, many of these materials have increased strength/weight ratios, enhanced conductivities, and improved optical or magnetic properties. These new properties make nanomaterial development so exciting and are the reason they hold the promise of great economic potential.

Nanomaterials are often classified by their physicochemical characteristics or structure. The four classes of materials of which nanoparticles are typically composed include elemental carbon, carbon compounds, metals or metal oxides, and ceramics. The nanometer form of metals, such as gold, and metal oxides, such as titanium dioxide, are the most common engineered nanomaterials being produced and used [Sellers et al. 2009]. Nano-sized silica, silver, and natural clays are also common materials in use. The carbon nanotube is a unique nanomaterial being investigated for a wide range of applications. These tubes are cylinders constructed of rolled-up graphene sheets. Another interesting carbon structure is a fullerene (also known as a Bucky Ball). These are spherical particles usually constructed from 60 carbon atoms arranged as 20 hexagons and 12 pentagons. As shown in Figure 1, the structure resembles a geodesic dome (designed by architect Buckminster Fuller, hence the name). Nanomaterials are widely used across industries and products, and they may be present in many forms.

Significant international health and safety research and guidance concerning the handling of nanomaterials is underway to support risk management of commercial developments. Both risks and rewards are inherent in these new materials. Scientists around the world are conducting toxicological studies on these nanomaterials, and initial findings are concerning. Animals exposed to titanium dioxide (TiO_2) and carbon nanotubes (CNTs) have displayed pulmonary inflammation [Chou et al. 2008; Rossi et al. 2010; Shvedova et al. 2005]. Other studies have shown that nanoparticles can translocate to the circulatory system and to the brain and cause oxidative stress [Elder et al. 2006; Wang et al. 2008]. Perhaps the most troubling finding is that CNTs can cause asbestos-like pathology in mice [Poland et al. 2008; Takagi et al. 2008].

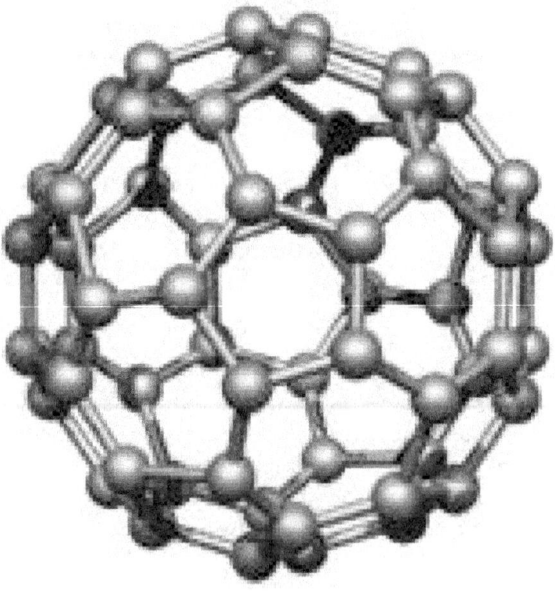

Figure 1. Atomic structure of a spherical fullerene

1.2 Industry Overview

In March 2006, the Woodrow Wilson International Center for Scholars created an inventory of 212 consumer products or product lines that incorporate nanomaterials (http://www.nanotechproject.org/inventories /consumer/analysis_draft/). These products were broken down into eight categories using a publically available consumer product classification system. As of March 2011, the number of consumer products has increased by 521% (212 to 1,317 nano-enabled products) with products coming from more than 24 nations [WWICS 2011]. These products include acne lotions, antimicrobial treatment for socks, sunscreens, food supplements, components for computer hardware (such as processors and video cards), appliance components, coatings, and hockey sticks. Of the current 1,317 nano-enabled products, the largest product category with 738 products was health and fitness. The most common type of nanomaterial used in these products was silver (313 products), followed by carbon (91 products) and titanium dioxide (59 products).

Roco [2005] reports that worldwide, the investment in nanotechnology has increased from $432 million in 1997 to about $4.1 billion in 2005. In this same time period, the U.S. government investment in nanotechnology has increased to nearly $1.1 billion. Estimates made in 2000 suggested that $1 trillion in products will use nanotechnology in some way by 2015. The National Science Foundation estimates that the number of workers in this industry will increase to 2 million worldwide by 2015.

Currently, most production facilities are relatively small, with lab, bench, or, at most, pilot plant operations [Genaidy et al. 2009]. This is also indicative of downstream users (applications and product development). As new manufacturing processes and technologies are developed and introduced, novel materials with unknown toxicological properties will require effective risk management approaches. As more of these products enter the market, concern about the health and safety of the workers grows.

1.3 Occupational Safety and Health Management Systems

Control measures for nanoparticles, dusts, and other hazards should be implemented within the context of a comprehensive occupational safety and health management system [ANSI/AIHA 2012]. The critical elements of an effective occupational safety and health management system include management commitment and employee involvement, worksite analysis, hazard prevention and control, and sufficient training for employees, supervisors, and managers (www.osha.gov/Publications/safety-health-management-systems.pdf). In developing measures to control occupational exposure to nanomaterials, it is important to remember that processing and manufacturing involve a wide range of hazards. Conducting a preliminary hazard assessment (PHA) encompasses a qualitative life cycle analysis of an entire operation, appropriate to the stage of development:

- Chemicals/materials being used in the process
- Production methods used during each stage of production
- Process equipment and engineering controls employed
- Worker's approach to performing job duties

- Exposure potential to the nanomaterials from the task/operations
- The facility that houses the operation

The steps taken to perform PHAs for specific operations should be documented to let others know what was done and to help others understand what works. PHAs are frequently conducted as initial risk assessments to determine whether more sophisticated analytical methods are needed and to prepare an inventory of hazards and control measures needed for these hazards. One or two individuals with a health and safety background and knowledge of the process can perform PHAs. As part of the assessment, the health and safety professional should evaluate the magnitude of the emissions (or potential emissions) and the effects of exposure to these emissions. PHAs are an important first step toward developing control measures that can be considered during the planning stage. Essentially, hazard control should be an integral component of facility, process, and equipment design and construction. This includes design for inherent process safety. The use of engineering controls should be considered as part of a comprehensive control strategy for hazards associated with processes/tasks that cannot be effectively eliminated, substituted for, or contained through process equipment modifications.

The standards for an occupational health and safety management system, as outlined in ANSI/AIHA Z10 [ANSI/AIHA 2012] and BSI 18001 [BSI 2007c], promote a continuous improvement cycle (plan, do, check, act), which does not have an exit point and is the basis for worksite analysis. Figure 2 illustrates how control measures are incorporated into an occupational safety and health management system. The continuous improvement loop is applicable to all hazards in a process/facility (e.g., airborne contaminant exposures, ergonomic, combustible dusts, fire safety, and physical hazards). The hazard assessment should be reviewed during each cycle described by Figure 2 and periodically updated when major changes occur. Although the optimal time to undertake a PHA is during the design stage, hazard assessments can also be done during the operation of a facility and have the benefit of using existing data.

After the PHA is complete, the nanomaterial risk management plan is designed to avoid or minimize hazards discovered during the assessment. The following options should be considered:

- Automated product transfer between operations. A process that allows for continuous process flow to avoid exposures caused by workers handling powdered or vaporous materials.
- Closed-system handling of powdered or vaporous materials, such as screw feeding or pneumatic conveying.
- Local exhaust ventilation. Steps should be taken to avoid having positive pressure ducts in work spaces because leakage from ducts can cause exposures. Ducts or pipes should be connected using flanges with gaskets that prevent leakage.
- Continuous bagging for the intermediate output from various processes and for final products. A process discharges material into a continuous bag that is sealed to eliminate dust exposures caused by powder handling. Bags are heat sealed after loading.

- Minimizing the container size for manual material handling. Minimizing the size of the container or using a long-handled tool is recommended so that the worker does not place his breathing zone inside the container (as shown in Figure 3). NIOSH recommends a maximum container depth of 25 inches [NIOSH 1997]. If large containers are required, engineering controls to provide a barrier between the container and the breathing zone of the worker are recommended.

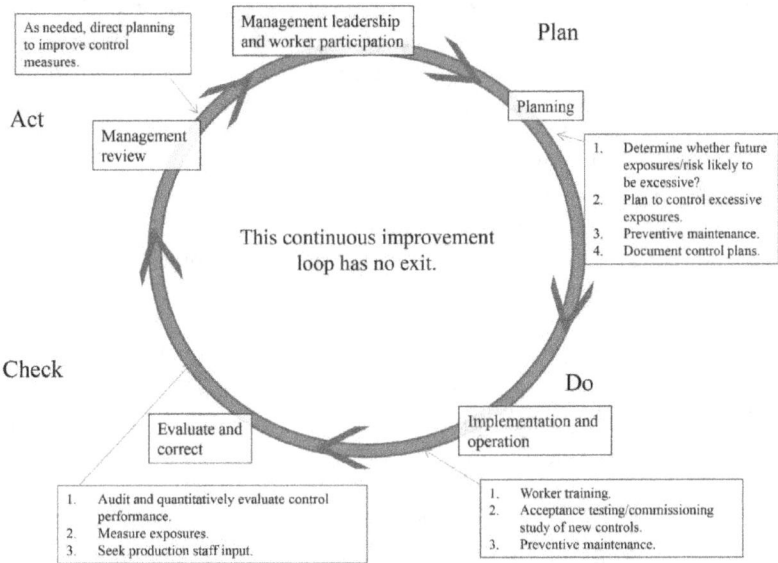

Figure 2. How control measures are selected, implemented, and managed into an occupational safety and health management system. (adopted from [ANSI 2005])

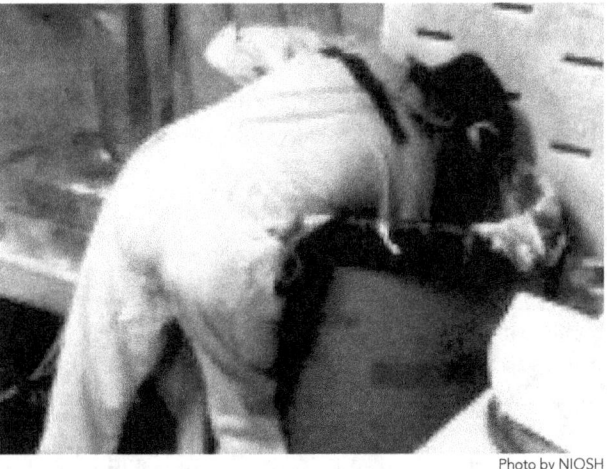

Figure 3. Worker reaching into drum

Many good resources are available on the occupational safety and health risk management of nanomaterials. Comprehensive documents have been produced by a number of organizations. Some of these are listed in Appendix A.

1.3.1 Prevention through Design (PtD)

The concept of Prevention through Design (PtD) is to design out or minimize hazards, preferably, early in the design process. PtD is also called inherent or intrinsic safety, safety by design, design for safety, and safe design. When PtD is implemented, the control hierarchy is applied by designers (e.g., engineers, architects, industrial designers) and business leaders (e.g., owners, purchasers, managers) who consider the benefits of designing safety into things external to the worker to prevent work-related injuries and illnesses.

PtD strategies, like the hierarchy of controls, can take many forms. Elimination and substitution measures are desirable, but these strategies may be difficult to implement when working with nanomaterials because these materials are likely being used for their unique properties. The pharmaceutical industry has addressed some of these challenges since their products must be contained rather than removed or eliminated from the process. They have adopted a containment hierarchy of controls that addresses designing inherent safety into the process [Brock 2009]. The initial levels of containment include elimination and substitution as well as product, process, and equipment modifications. Only after efforts have been made to design the process to reduce potential emissions sources should engineering controls be considered.

Other PtD strategies can be considered:

- Limiting process inventories by producing the nanomaterials as they are consumed in the process.
- Operating a process at a lower energy state (e.g., lower temperature or pressure), which typically results in lower fugitive emissions and therefore safer operation.
- Using fail-safe devices where possible. Fail-safe devices are designed so that if they fail, the system reverts to a safer condition. An example of a fail-safe device is a valve controlling a reagent for a reaction. If the safe condition for the system is for this valve to be closed, the fail-safe valve would automatically close in the event of a failure.
- Installing a closed transport system to eliminate worker exposures during transport activities.

PtD strategies typically do not include administrative controls and personal protective equipment (PPE) as the primary controls. These measures require worker interaction with the process or active steps to limit the extent of the hazard. Most effective PtD approaches reduce or eliminate hazardous conditions without relying on input from workers. Humans are generally recognized as being much less reliable than most machines, particularly in emergencies [Kletz 2001]. The use of administrative controls and PPE in PtD strategies is generally for redundancy—further safeguards should the primary control fail.

The ideal time to develop a PtD strategy is during the development phase of a process, material, or facility. As the nanotechnology field is still in its relative infancy, there are numerous opportunities to implement PtD in the early stages. The manner in which these materials are handled and processed can largely affect the overall safety of the process, and the health and safety of workers may be significantly improved through the implementation of a PtD strategy.

1.3.2 OELs as Applied to Nanotechnology

Occupational exposure limits (OELs) are useful in reducing work-related health risks by providing a quantitative guideline and basis to assess the worker exposure potential and the performance of engineering controls and other risk management approaches. Currently, no regulatory standards for nanomaterials have been established in the United States. However, NIOSH has recently published two current intelligence bulletins (CIBs) regarding occupational exposures to nanomaterials. In a CIB on titanium dioxide (TiO_2), NIOSH recommends exposure limits of 2.4 mg/m^3 for fine TiO_2 and 0.3 mg/m^3 for ultrafine (including engineered nanoscale) TiO_2, as time-weighted average (TWA) concentrations for up to 10 hours per day during a 40-hour work week [NIOSH 2011]. In a CIB on carbon nanotubes and nanofibers, NIOSH recommends that worker exposure be limited to no more than 1 µg/m3 [NIOSH 2013].

Other countries have established OELs for various nanomaterials. For example, the British Standards Institute recommends working exposure limits for nanomaterials based on various classifications such as solubility, shape, and potential health concerns as related to larger particles of the same substance [BSI 2007b]. Germany's Institut für Arbeitsschutz der Deutschen Gesetzlichen Unfallversicherung, an institute for worker safety, has published similar guidelines [IFA 2009].

In the absence of governmental or consensus guidance on exposure limits, some manufacturers have developed suggested OELs for their products. For example, Bayer has established an OEL of 0.05 mg/m3 for Baytubes® (multiwalled CNTs) [Bayer MaterialScience 2010]. For Nanocyl CNTs, the no-effect concentration in air was estimated to be 2.5 µg/m^3 for an 8-hr/day exposure [Nanocyl 2009].

Another approach that may be taken when OELs are absent is the ALARA concept, As Low As Reasonably Achievable. While ALARA is generally the goal for all occupational exposures, this concept is particularly useful when OELs are absent or in the case of contaminants with unknown toxicity.

1.3.3 Control Banding

Control banding is a qualitative risk characterization and management strategy, intended to protect the safety and health of workers in the absence of chemical and workplace standards. Control banding groups workplace risks into hazard bands based on evaluations of hazard and exposure information [NIOSH 2009b]. Note that control banding is not intended to be a substitute for OELs and does not alleviate the need for environmental monitoring or industrial hygiene expertise.

To determine the appropriate control scheme, one should consider the characteristics of the substance, the potential for exposure, and the hazard associated with the substance. Four main control bands, based on an overall risk level, have been developed:

- Good industrial hygiene (IH) practice, general ventilation, and good work practices
- Engineering controls including fume hoods or local exhaust ventilation
- Containment or process enclosure allowing for limited breaks in containment
- Special circumstances requiring expert advice

One basic principle of control banding is the need for a method that will return consistent, accurate results even when performed by nonexperts. Other requirements include having a user friendly strategy, readily available required information (e.g., material safety data sheet [MSDS]), practical guidance on applying the strategy, and worker confidence in the results. With the absence of OELs, control banding can be a useful approach in the risk management of nanomaterials [Maynard 2007; Schulte et al. 2008; Thomas et al. 2006; Warheit et al. 2007]. Several control banding tools are available for use with engineered nanomaterials. The CB Nanotool, for example, bases the control band for a particular task on the overall risk level, which is determined by a matrix that uses severity scores and probability scores [Paik et al. 2008]. The severity score is based on the toxicological effects of the nanomaterial, while the probability score relates to the potential for employee exposure. The health hazard categories for some control banding approaches are based upon the European Union risk phrases, while exposure potentials include the volume of the chemical used and the likelihood of airborne materials, estimated by the dustiness or volatility of the source compound [Maidment 1998].

CHAPTER 2
Exposure Control Strategies and the Hierarchy of Controls

Controlling exposures to occupational hazards is the fundamental method of protecting workers. Traditionally, a hierarchy of controls has been used as a means of determining how to implement feasible and effective controls. Figure 4 shows one representation of this hierarchy. The idea behind the hierarchy of controls is that the control methods at the top of the triangle are generally more effective in reducing the risk associated with a hazard than those at the bottom. Following the hierarchy normally leads to the implementation of inherently safer systems, ones where the risk of illness or injury has been substantially reduced. Designing out hazards early in the design process is a basic tenet of PtD. When PtD is implemented, the control hierarchy is applied by designers and owners/managers to include safety into the process.

The following sections discuss each element of the hierarchy of controls—elimination, substitution, engineering controls, administrative controls, and PPE— and how it may relate to nanotechnology.

2.1 Elimination

Elimination and substitution are generally most cost effective if implemented when a process is in the design or development stage. If done early enough, implementation is simple and, in the long run, can result in substantial savings (e.g., cost of protective equipment, first cost and operational cost for ventilation system). For an existing process, elimination or substitution may require major changes in equipment and/or procedures in order to reduce a hazard.

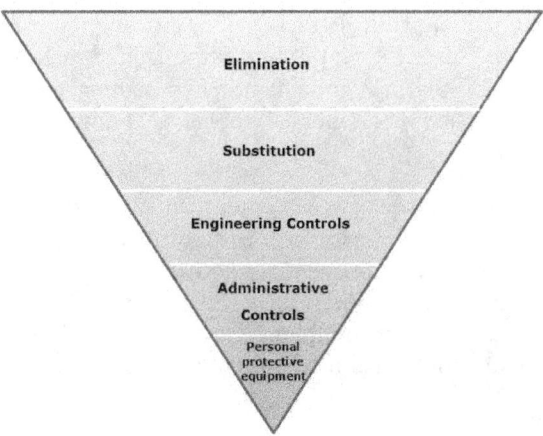

Figure 4. Graphical representation of the hierarchy of controls

Elimination is the most desirable approach in the hierarchy of controls. As its name implies, the idea behind elimination is to remove the hazard. Eliminating hazards is generally easiest to accomplish at the design stage, while the material, process, and/or facility is being developed. An example of elimination in a process step might be the removal of an incoming inspection step for nanomaterials. An incoming inspection that requires opening a package containing nanomaterials leads to the potential of aerosolization of those materials and therefore a potential hazard to the inspector. Eliminating the inspection step removes the hazard, thus creating an inherently safer process.

2.2 Substitution

Within the hierarchy of controls, the purpose of substitution is to replace one set of conditions having a high hazard level with a different set of conditions having a lower hazard level. Examples of substitution could include replacing a solvent-based (i.e., flammable) material with a water-based material, substituting a highly toxic material for one of lower toxicity, or changing a process's operating conditions so they are less severe (e.g., reduced pressure). Substitution of a nanomaterial may be difficult since it was likely introduced for its particular properties; however, some substitution may be possible. Substituting a nanomaterial slurry for a dry powder version will reduce aerosolization and provide a level of protection for workers handling the material. The specific nanomaterial should also be assessed because in some cases a less hazardous nanomaterial may provide the desired performance.

2.3 Engineering Controls

Engineering controls protect workers by removing hazardous conditions (e.g., local exhaust ventilation that captures and removes airborne emissions) or placing a barrier between the worker and the hazard (e.g., isolators and machine guards). Well-designed engineering controls can be highly effective in protecting workers and will typically be passive, that is, independent of worker interactions. It is important to design engineering controls that do not interfere with the productivity and ease of processing for the worker. If engineering controls make the operation more difficult, there will be a strong motivation by the operator to circumvent these controls. Ideally, engineering controls should make the operation easier to perform rather than more difficult. A good mantra in designing engineering controls is to "make it easier to do it the safe way." This also applies to administrative controls that are discussed later.

The initial cost of engineering controls can be higher than administrative controls or personal protective equipment (PPE); however, over the long term, operating costs are frequently lower and, in some instances, can provide a cost savings in other areas of the process. The major benefit of engineering controls over administrative controls or PPE is, however, the inherent safety of the worker under a variety of conditions and stress levels. The use of engineering controls reduces the potential for worker behavior to impact exposure levels.

Thus, when elimination and substitution are not viable options, the most desirable alternative for mitigating occupational hazards is to employ engineering controls. Engineering controls are likely the most effective and applicable control strategy for most nanomaterial processes.

In most cases, they should be more feasible than elimination or substitution and, given the potential toxicity of many nanomaterials, should prove to be more protective than administrative controls and PPE.

Engineering controls are divided into two broad categories for discussion below: ventilation and nonventilation controls.

2.3.1 Ventilation

The general concept behind ventilation for controlling occupational exposures to air contaminants, including nanomaterials, is to remove contaminated air from the work environment. The efficiency of the ventilation system can be affected by its configuration and flow volumes of both the air supplied to and the air exhausted from the work space. Effective ventilation applies to a wide range of applications including office heating, ventilating, and air conditioning (HVAC); infection control in healthcare; and control of emissions in industrial processes. Ventilation for occupant comfort, HVAC, is a specialized application of dilution ventilation and is not within the scope of this document. Filtration is a topic directly affecting ventilation; exhaust air laden with nanomaterials may need to be cleaned before being released into the environment.

General ventilation can be used to achieve several goals for workplace contaminant control. A properly designed supply air ventilation system can provide plant ventilation, building pressurization, and exhaust air replacement. As new local exhaust hoods are installed in the production area, it is important to consider the need for replacement air, the location of the hood installation, and the need to rebalance the ventilation system. In general, it is necessary to balance the amount of exhausted air with a nearly equal amount of supply air. Without this replacement air, uncontrolled drafts will occur at doors, windows, and other openings; doors will become difficult to open due to the high pressure difference, and exhaust fan performance may degrade. In addition, turbulence created through high pressure differentials can defeat the design intent of the ventilation. Placement of the air supply registers in relation to other exhaust ventilation systems is important so that they do not negatively impact the desired performance. The use of general ventilation for dilution of contaminants being generated in the space should be restricted in its use depending on several factors discussed below.

General ventilation used for dilution of contaminants by its nature is inefficient. One of two methods, recirculated air or single-pass air, may be used for this purpose. As the terms imply, recirculated air involves the treatment of exhaust air prior to its being returned to the area from which it was exhausted. Single-pass air is exhausted to the outside and may or may not require treatment prior to discharge. Both of these methods are expensive—the treatment of the recirculated air involves both first-cost and operating-cost penalties, while makeup-air treatment for single-pass air is inherently costly.

According to the American Conference of Governmental Industrial Hygienists (ACGIH) *Industrial Ventilation: A Manual of Recommended Practice for Design* (hereafter referred to as the *Industrial Ventilation Manual*), dilution ventilation (i.e., air changes) to control exposure should be used only under specific conditions. Dilution ventilation for controlling health hazards is restricted by four limiting factors: (1) the quantity of contaminant generated must not be too great or the airflow rate necessary for dilution will be impractical, (2) workers must be far enough away from the contaminant source or the evolution of contaminant must be in sufficiently low concentrations so that workers will not have an exposure in excess of the established threshold limit values (TLV®), (3) the toxicity of the contaminant must be low, and (4) the evolution of contaminants must be reasonably uniform [ACGIH 2013]. There are several issues with using dilution ventilation to control nanomaterial concentrations, including (1) there are no occupational exposure limits (TLVs mentioned above) or health effects data for many of the nanomaterials, (2) the toxicology data from some nanomaterials indicate that they may be associated with adverse health effects, and (3) it is difficult or impossible to calculate proper air change rates for contaminant control due to the variability in most operations. Therefore, local exhaust ventilation and good work practices should be used for controlling exposure, and air change rates should be based on the heat load requirements, general air movement, and comfort needs.

The use of supply air for maintaining proper pressurization between production and nonproduction areas is a reasonable approach to reducing the exposure to nanomaterials outside of the immediate work zone. The fugitive emissions from nanomaterial production and processing may result in high background concentrations in the production area. When adjacent plant areas are nonproduction areas (e.g., office, quality assurance/control labs) or production areas where nanomaterials are not used, infiltration of nanoparticles may occur and result in the exposure of workers in those areas. Therefore, a negative air pressure differential should be maintained in the nanomaterial production area with respect to adjacent rooms/areas. This will help reduce the potential migration of airborne nanomaterials and exposure to other workers in adjacent rooms or areas. To maintain a slight negative pressure, the room supply air volume should be slightly less than the exhaust air. A general guide is to set a 5% flow difference between supply and exhaust flow rates but no less than 50 cfm [ACGIH 2013]. As with any good engineering control, a real-time monitor of differential pressure between areas should be employed, preferably with the control capability to modify airflows to maintain the required pressure differential.

2.3.1.1 Local Exhaust Ventilation

Local exhaust ventilation (LEV) is the application of an exhaust system at or near the source of contamination. If properly designed, it will be much more efficient at removing contaminants than dilution ventilation, requiring lower exhaust volumes, less make-up air, and, in many cases, lower costs. By applying exhaust at the source, contaminants are removed before they get into the general work environment. When designing a local exhaust ventilation system, it is important to understand the transport mechanisms of the contaminants that are to be removed. This will allow the design to use optimal flow rates and capture locations, maximizing the contaminant capture while minimizing impact on the process and reducing operating costs. LEV typically involves five components [Washington State L & I, no date]:

- Exhaust hood. Examples include an enclosing hood to contain the contaminant, a receiving hood to capture or receive a contaminant that is released at a high velocity (e.g., grinding swarf), or simply an open duct.

- Duct. Transports the contaminant through the exhaust ventilation system.
- Air cleaner. Reduces the concentration of the contaminant in the exhaust air stream; may or may not be required.
- Fan. Moves the air through the exhaust system.
- Exhaust stack. Installed where the exhaust system discharges the air.

The **exhaust hood** captures the contaminant released by the process. It should be designed for the specific process being controlled, an important consideration for hot processes and those processes generating contaminants at high velocities. In either case, induced air flow (from high velocity air streams or rising air from a hot process) can overwhelm an insufficiently designed hood and allow contaminants to escape into the work environment. An important hood design factor is the capture velocity. This is the velocity of air needed to overcome contaminant velocity as well as room air currents. ACGIH *Industrial Ventilation Manual* contains a large collection of industrial ventilation hood designs for a wide selection of industrial processes [ACGIH 2013]. Though many of these designs have not been tested with nanomaterials, most are expected to perform effectively with these materials. An important consideration in hood design with nanomaterials is to provide the appropriate flow rates to prevent fugitive emissions without causing conditions that will remove nanomaterials from the process stream. Because of their very low mass, entrainment of nanomaterials in airflow streams occurs much more readily than with higher-mass particles.

Duct systems transport air between the various components of the LEV system. Designing duct systems requires balancing several factors. Duct losses caused by friction will increase with higher duct velocities, resulting in increased fan requirements and higher energy consumption; however, using larger ducts (in an effort to reduce duct velocity) results in increased duct purchase costs. A detailed method for designing and sizing LEV duct systems is provided by ACGIH [ACGIH 2013]. The choice of duct materials and sealing methods is particularly important when dealing with nanomaterials. The duct material needs to be impervious to the nanomaterials and suitable for use with nanomaterials having increased reactivity. The joints in the ducts should be sealed in such a way as to contain the nanomaterials.

Fans move air throughout the LEV system. Fans need to be sized to ensure adequate air flow while overcoming the system pressure drop (i.e., resistance to flow). Pressure drop is encountered when air is accelerated, such as within a hood; through ductwork due to frictional losses, particularly in fittings such as elbows; and through filters and other air-cleaning devices. Fan selection affects not only the control effectiveness of the LEV system but also its energy consumption. The fan system and the make-up air conditioning are typically the two greatest energy-consuming components of an LEV system. Proper fan selection needs to balance both control performance and operating efficiency [ACGIH 2013]. The same leakage and reactivity factors mentioned in the section on ductwork apply to fan selection.

Air cleaning is an important component of the LEV system, particularly if the exhaust air is returned to the building environment. Air cleaning involves the removal of gases and vapors, often with scrubbers and sorbent systems; however, in the case of nanomaterials, particulate removal systems will be required to eliminate them from the air stream. Cyclones, scrubbers, and other similar systems can be used to remove larger-sized particles, but smaller, nanoparticles will most likely be collected by filtration (see next section, Air Filtration).

2.3.1.2 Air Filtration

Air filtration removes unwanted particulate from an air stream. Particulate air filters are classified as either mechanical or electrostatic filters. Although the two types of filters have important performance differences between them, both are fibrous media or membranes and are used extensively in HVAC and industrial applications. Efficiency is dependent on several factors including fiber diameters, packing density, and material used. A fibrous filter is an assembly of fibers that are randomly laid perpendicular to the airflow. The fibers may range in size from less than 1 μm to greater than 50 μm in diameter. Filter packing density ranges from 1%–30%. Fibers are made from cotton, fiberglass, polyester, polypropylene, or a number of other materials [Davies 1977].

Fibrous filters of different designs are used for various applications. Three types are used for capturing particulate:

- Flat-panel filters contain all the media in the same plane. This design keeps the filter face velocity and the filter media velocity roughly the same.
- Pleated filters have additional filter media added to reduce the air velocity through the filter. This allows for an increased collection efficiency for a given pressure drop. Alternatively, pleated filters can be used to reduce the pressure drop for a given airflow velocity because of the larger filter area.
- Pocket or baghouse filters allow the flow of exhaust air through small pockets or bags consisting of filter media. As with pleated filters, the increased surface area of the pocket filter reduces the velocity of the airflow through the filter media, allowing increased collection efficiency for small particles at a given pressure drop.

Figure 5 presents four different collection mechanisms that govern particulate air filter performance:

- Diffusion is the result of the random (Brownian) motion of a particle. The particle may contact a fiber on its path through the filter.
- Interception occurs when the radius of a particle moving along an air streamline is greater than the distance from the streamline to the surface, thus causing the particle surface to contact the surface of the fiber. The particle adheres to the fiber due to intermolecular forces.
- Inertial impaction occurs when an air stream bends around a fiber, and a particle traveling in that air stream continues in a straight path due to particle inertia. The particle collides with the fiber and adheres to it due to intermolecular forces.
- Electrostatic attraction occurs when the particle and the fiber are oppositely charged. As the force of this attraction is governed by the charge-to-mass ratio of the particle, it becomes more effective as particle size decreases.

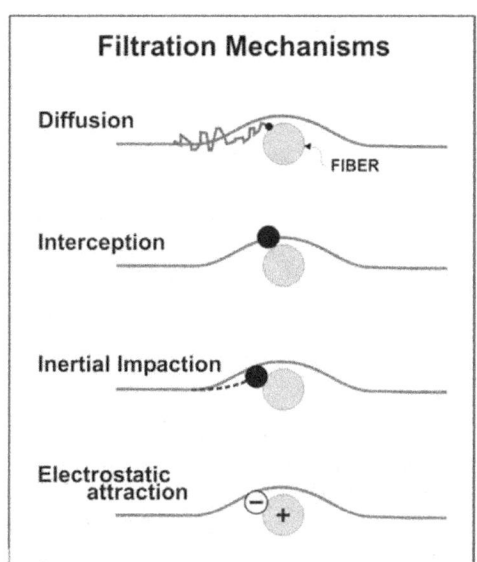

Figure 5. Four primary filter collection mechanisms

These mechanisms apply mainly to mechanical filters and are influenced by particle size. Impaction and interception are the dominant collection mechanisms for particles greater than 0.2 μm, and diffusion and electrostatic attraction are dominant for particles less than 0.2 μm, including nanomaterials. The combined effect of these collection mechanisms results in the classic collection efficiency curve, shown in Figure 6.

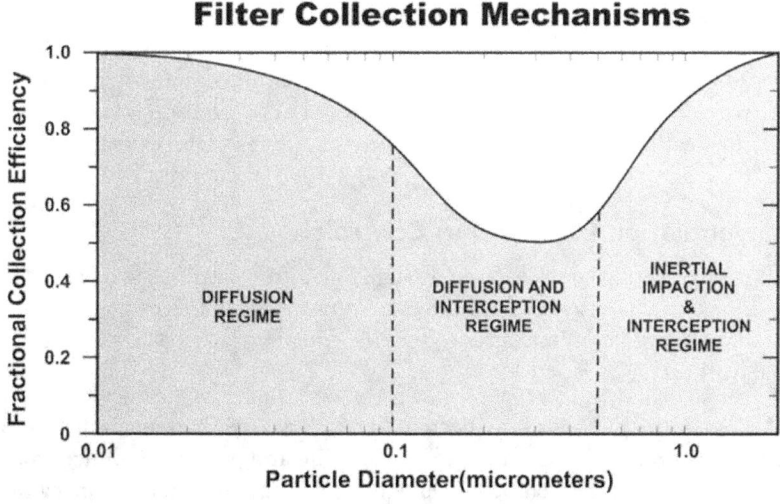

Figure 6. Collection efficiency curve, i.e., fractional collection efficiency versus particle diameter for a typical filter (Used with permission from Lee and Liu [1980].)

Research on common air filter materials has shown that fractional efficiency for collection of particles of different sizes is consistent with the single fiber theory [Heim et al. 2005; Kim et al. 2007; Shin et al. 2008]. Kim et al. [2006] found that humidity has little effect on particle collection efficiency. Huang et al. [2007b] determined that the use of electrostatic filters (commonly used for respirators) improves particle collection in the 0.1–1-μm particle size range. Testing of respirator filters showed that the most penetrating particle size (MPPS) shifted from 30–60 nm to 200–300 nm following treatment of respirators by liquid isopropanol, which removes electrostatic charges on the filter materials [Rengasamy et al. 2009]. This result suggests that capture by electrostatic forces is important for particles in the 250–300-μm range. Overall, filters appear to behave in a manner consistent with theoretical predictions that common filter materials allow for efficient collection through diffusion of nanoparticles less than about 10 nm [Heim et al. 2005; Huang et al. 2007b; Kim et al. 2007; Shin et al. 2008].

Some researchers have found evidence of thermal rebound, which increases particle penetration through filters for nanoparticles in the size range of 1–10 nm [Bałazy et al. 2004; Kim et al. 2006]; however, several other filter testing studies did not reveal this effect, even at higher temperatures [Heim et al. 2005; Huang et al. 2007b; Kim et al. 2007; Shin et al. 2008]. The thermal rebound effect is a result of the thermal velocity of the particle exceeding the critical sticking velocity for a particle on a filter, allowing the particle to move past the filter fiber and penetrate the filter. The critical sticking velocity of an incident particle is defined as the maximum impact speed at which the particle will stick to a surface; above this velocity, the particle will bounce and not stick to the filter. The primary adhesive forces for nanomater-sized particles are the London-van der Waals forces. These forces are caused by random movement of electrons creating complementary dipoles between particle and filter material [Hinds 1999]. As the particle gets smaller, it is more difficult to remove the particle from surfaces.

High efficiency particulate air (HEPA) filtration is commonly used for applications requiring reliably high filtration. By definition, HEPA filters are 99.97% efficient at the most penetrating particle size of 0.3 microns (Figure 6). These filters are disposable and are usually replaced when the pressure drop exceeds a predetermined number, typically 100 mm water gauge (wg). When properly sized and installed, HEPA filtration is appropriate for nanomaterial applications both for ventilation systems and respiratory protection.

2.3.2 Nonventilation Engineering Controls

Nonventilation engineering controls cover a range of control measures (e.g., guards and barricades, material treatment, or additives). Nonventilation controls can be used in conjunction with ventilation measures to provide an enhanced level of protection for nanomaterial workers.

A variety of dust control methods have been used and evaluated in many industries and may be applicable to the processes used in the manufacturing and processing of nanomaterials [Smandych et al. 1998]. These methods include the enclosure of material-conveying equipment, such as belt and screw conveyers, as well as the use of pneumatic conveyance systems. Other work practices have been used to reduce the aerosolization of dust during bag filling, including minimizing leak paths by securing the bag to the outlet spout and wetting the outside of the bag to prevent surface dust from becoming airborne. Research over the years in a variety of industrial settings has shown that water spray application is effective in lowering respirable dust levels [Mukherjee et al. 1986]. The use of atomization nozzles was shown to be one of the most effective water-spray delivery systems in dust knockdown performance tests. Water sprays lower respirable dust concentrations by knocking down the dust, fibers, and particles, and they also can induce airflow to direct the remaining dust away from the workers.

Other nonventilation engineering controls include many devices developed for the pharmaceutical industry, including isolation containment systems [Hirst et al. 2002]. One of the most common flexible isolation systems is glove box containment, which can be used as an enclosure around small-scale powder processes, such as mixing and drying. Rigid glove box isolation units also provide a method for isolating the worker from the process and are often used for medium-scale operations involving transfer of powders. Glove bags are similar to rigid glove boxes, but they are flexible and disposable. They are used for small operations for containment or protection from contamination. Another nonventilation control used in this industry is the continuous liner system, which allows the filling of product containers while enclosing the material in a polypropylene bag. This system is often used for off-loading materials when the powders are to be packed into drums.

2.4 Administrative Controls

Administrative controls and PPE are frequently used with existing processes where hazards are not well controlled. This could occur when engineering control measures are not feasible or do not reduce exposures to an acceptable level. Administrative controls (which include training, job rotation, work scheduling, and other strategies to reduce exposure) and PPE programs may be less expensive to establish but, over the long term, can be very costly to sustain. These methods for protecting workers have also proven to be less effective than other measures and often require significant effort by the affected workers [ACGIH 2013; DiNardi 2003]. A valuable application of administrative controls is as a redundancy to engineering controls. While the engineering controls provide the primary protection for the worker, the administrative controls serve as back-up should the engineering control fail.

NIOSH recommends that facilities implement the following work practices as part of an overall strategy to control worker exposure to nanomaterials: (1) Educate workers on the safe handling of engineered nanomaterials to minimize the likelihood of inhalation exposure and skin contact. (2) Provide information on the hazardous properties of the materials being handled with instructions on how to prevent exposure. (3) Encourage workers to use hand-washing facilities before eating, smoking, or leaving the worksite. (4) Provide additional control measures (e.g., use of a buffer area, decontamination facilities for workers if warranted by the hazard) to ensure that engineered nanomaterials are not transported outside of the work area. (5) Where there is the potential for area or personnel contamination, provide facilities for showering and changing clothes to prevent the inadvertent contamination of other areas (including take-home) caused by the transfer of nanomaterials on clothing and skin. (6) Avoid handling nanomaterials in the open air in a "free particle" state. (7)

Store dispersible nanomaterials, whether suspended in liquids or in a dry particle form, in closed (tightly sealed) containers whenever possible. (8) Ensure work areas and designated equipment (e.g., balance) are cleaned at the end of each work shift, at a minimum, using either a HEPA-filtered vacuum cleaner or wet wiping methods (where the use of liquid does not create additional safety hazards). Dry sweeping (i.e., using a broom) or compressed air should not be used to clean work areas. Cleanup should be conducted in a manner that prevents worker contact with wastes. (9) Dispose of all waste material in compliance with all applicable federal, state, and local regulations. (10) Avoid storing and consuming food or beverages in workplaces where nanomaterials are handled [NIOSH 2009a].

2.5 Personal Protective Equipment (PPE)

PPE (e.g., respirators, gloves, protective clothing) is the least desired option for controlling worker exposures to hazardous substances. PPE is used when engineering and administrative controls are not feasible or effective in reducing exposures to acceptable levels or while controls are being implemented. It is the last line of defense after engineering controls, work practices, and administrative controls. A program that addresses the hazards present, employee training, and PPE selection, use, and maintenance should be in place when PPE is used.

2.5.1 Skin Protection

Nanomaterials have been shown to accumulate in hair follicles, and quantum dots have been shown to penetrate the skin into the dermis [Smijs and Bouwstra 2010]. Flexing the skin may enhance skin penetration [Smijs and Bouwstra 2010; Tinkle et al. 2003]. Woskie [2010] recommends wearing gloves, gauntlets, and laboratory clothing or coats when working with nanoparticles. Other studies of specifically engineered nanomaterials have resulted in the material not penetrating beyond the stratum corneum. Of importance is to establish a barrier between the potentially hazardous material and the skin.

Air-tight polyethylene was found to be more resistant to nanoparticle penetration by diffusion than cotton or polyester; gloves made of latex, neoprene, or nitrile resisted nanoparticle penetration "during exposure of a few minutes" [Woskie 2010]. Proper selection of gloves should take into account the resistance of the glove to the nanomaterial and any other chemicals or liquids with which the hands may come into contact. Gloves should be changed whenever they show visible signs of wear or contamination. Gao et al. [2011] studied nano- and submicron-size (30–500 nm) iron oxide particle penetration through some protective clothing materials. They found that particle penetration increased with increasing wind velocity and increasing particle size. Results from the study indicated that the MPPS for protective clothing materials tested was found to be about 300 to 500 nm, compared to an MPPS for N-95 respirators of 50 nm.

2.5.2 Respiratory Protection

Respiratory protection is used to reduce worker exposures to acceptable levels in the absence of effective engineering controls, during the installation or maintenance of engineering controls, for short-duration tasks that make engineering controls impractical, and during emergencies. The decision to use respiratory protection should be based upon professional judgment, hazard assessment, and risk management practices to keep worker inhalation exposures below an internal control or an exposure limit. Several types of NIOSH-certified respirators (e.g., disposable filtering facepiece, half-mask elastomeric, full facepiece, powered, airline, self-contained) can provide different levels of expected protection to airborne particulate when used in the context of a complete respirator program [60 Fed. Reg. 30336 (1995); NIOSH 2004]. In a survey designed to better understand health and safety practices in the carbonaceous nanomaterial industry, NIOSH found half-mask elastomeric particulate respirators fitted with HEPA filtration media to be the most commonly used respiratory protection, followed by disposable filtering facepiece respirators [Dahm et al. 2011].

The 2009 NIOSH Approaches to Safe Nanotechnology document as well as the Current Intelligence Bulletins on titanium dioxide and carbon nanotubes contain recommendations on respirator use and selection when working with nanoparticles. Recommendations from other organizations and a discussion of the scientific rationale for respirator selection have been reviewed [Shaffer and Rengasamy 2009]. Current respirator performance research suggests that NIOSH's traditional respirator selection tools apply to nanoparticles. NIOSH-certified respirators should provide the expected levels of protection, consistent with their assigned protection factor, and should be selected according to the NIOSH Respirator Selection Logic [NIOSH 2004] by the person who is in charge of the program and knowledgeable about the workplace and the limitations associated with each type of respirator. As part of the risk assessment process, respirators with 95-, 99-, or 100-class filters can be selected for workplaces with concentrations of nanoparticles near their MPPS (50 to 100 nm). Furthermore, NIOSH recommends that all elements of the OSHA Respiratory Protection Standard [29 CFR 1910.134] for both voluntary and required respirator use should be followed [63 Fed. Reg. 1152 (1998)].

Selection of respiratory protection for airborne particulate contaminants is typically done by dividing the measured or anticipated time-weighted average concentration of the airborne contaminant by the OEL and comparing that quotient to the respirator's assigned protection factor (APF). Alternatively, the respirator's APF can be multiplied by the OEL to find its maximum use concentration (MUC). The MUC is then compared to the TWA to select the appropriate respirator. In the absence of an OEL for nanoparticles, Woskie [2010] recommends that a health and safety professional "familiar with the workplace" choose the appropriate respirator based on goals for nanoparticle control, sampling results, and the capabilities of each type of respirator.

The NIOSH respirator selection logic recommends (and it is mandated by OSHA where the use of respirators is required) that respirators in the workplace be used as part of a comprehensive respiratory protection program. The program should include written standard operating procedures; workplace monitoring; hazard-based selection; fit-testing and training of the user; procedures for cleaning, disinfection, maintenance, and storage of reusable respirators; respirator inspection and program evaluation; medical qualification of the user; and the use of NIOSH-certified respirators [NIOSH 2004].

Several studies have been conducted of respirator media filtration performance against nanoparticles. Many employers provide filtering facepiece respirators (FFRs) due to their common availability and low cost. One study of N95 FFRs showed penetration levels by nanoparticles in the size range of ~30 to 70 nm, which exceeded the 5% level allowed by NIOSH [Balazy et al. 2006]. A later study used two test methods (challenges using a monodisperse aerosol and a polydisperse aerosol similar to the NIOSH certification test) and compared particle penetration of N-95 FFRs [Rengasamy et al. 2007]. Those authors found that a monodisperse aerosol challenge test using particles from 20 nm to 400 nm in diameter resulted in a MPPS near 40 nm. The monodisperse test found that two respirators exceeded the NIOSH 5% allowed penetration, but the exceedance was not statistically significant, while the polydisperse challenge produced penetration levels from 0.61% to 1.24%. The NIOSH-allowed penetration level of < 0.03% was not exceeded by P100 FFRs, but two nanoparticle test aerosols did exceed 1% penetration for two N99 FFRs [Eninger et al. 2008; Rengasamy et al. 2009]. Five models of N95 and two models of P100 FFRs challenged with 4–30-nm monodisperse aerosols provided approved levels of protection [Rengasamy et al. 2008]. Rengasamy et al. [2009] tested two models each of N95 and P100 respirators with monodisperse aerosols in the 4–30-nm range and the 20–400-nm range. The penetration levels were less than the NIOSH-allowed levels of < 5% and < 0.03% across all test methods used. The penetration was < 4.28% for the N95 respirators and < 0.009% for the P100 respirators at the MPPS range of 30–60 nm.

NIOSH-certified FFRs have been shown to provide "expected levels of filtration efficiency against polydisperse and monodisperse aerosols > 20 nm in size" [Rengasamy and Eimer 2011]. A study showed that eight commercially purchased models of NIOSH-approved N95 and P100 and CE-marked FFR models "provided expected levels of laboratory performance against nanoparticles" [Rengasamy et al. 2009].

CHAPTER 3
Nanotechnology Processes and Engineering Controls

3.1 Primary Nanotechnology Production and Downstream Processes

Currently, nanomaterials are produced using a variety of methods that provide conditions for the formation of desired shapes, sizes, and chemical composition. These production processes can be separated into six categories [HSE 2004; NNI, no date]:

- **Gas phase processes, including flame pyrolysis, high-temperature evaporation, and plasma synthesis.** This process involves the growth of nanoparticles by homogenous nucleation of supersaturated vapor. Nanoparticles are formed in a reactor at high temperatures when source material in solid, liquid, or gaseous form is injected into the reactor. These precursors are supersaturated by expansion and cooled prior to the initiation of nucleated growth. The size and composition of the final materials depend on the materials used and process parameters.

- **Chemical vapor deposition (CVD).** This process has been used to deposit thin films of silicon on semiconductor wafers. The chemical vapor is formed in a reactor by pyrolysis, reduction, oxidation, and nitridation and deposited as a film with the nucleation of a few atoms that coalesce into a continuous film. This process has been used to produce many nanomaterials including TiO_2, zinc oxide, silicon carbide, and, possibly most importantly, CNTs. The use of fluidized bed technology has been adopted as a way to prepare CNTs on a large scale at low cost [Wang et al. 2002]. This technology fluidizes CNT agglomerates and produces high yields necessary for larger-scale operations.

- **Colloidal or liquid phase methods.** Chemical reactions in solvents lead to the formation of colloids. Solutions of different ions are mixed to produce insoluble precipitates. This method is a fairly simple and inexpensive way to produce nanoparticles and is often used for the synthesis of metals (e.g., gold, silver). These nanomaterials may remain in liquid suspension or may be processed into dry powder materials often by spray drying and collection through filtration.

- **Mechanical processes including grinding, milling, and alloying.** These processes create nanomaterials by a "top-down" method that reduces the size of larger bulk materials through the application of energy to break materials into smaller and smaller particles. This technique has been referred to as nanosizing or ultrafine grinding.

- **Atomic and molecular beam epitaxy.** Atomic layer epitaxy is the process of depositing monolayers (i.e., layers one molecule thick) of alternating materials and is commonly used in semiconductor fabrication. Molecular beam epitaxy is another process for depositing highly controlled crystalline layers onto a substrate.

- **Dip pen lithography.** A "bottom-up" method is a production process that involves depositing a chemical on the surface of a substrate using the tip of an atomic force microscope (AFM). The AFM tips are coated with the chemical, which is directly deposited on a substrate in a specific pattern.

Downstream processes use engineered nanomaterials for product application and development. Examples of these tasks or operations include weighing, dispersion/sonication, mixing, compounding/extrusion, electro-spinning, packaging, and maintenance. These activities should be evaluated for potential sources of exposure.

3.2 Engineering Control Approaches to Reducing Exposures

Engineering controls are used to remove a hazard or place a barrier between the worker and the hazard, and though costs of engineering controls may be higher than that of administrative controls or PPE initially, over the long term, operating costs are often lower. A major advantage of engineering controls is that, when properly designed, they require little or no user effort or training to be effective. Many industries have implemented engineering controls to reduce exposure and risk of disease among their workers. The pharmaceutical industry uses hazardous (i.e., biologically active) liquids and powders that often do not have OELs. To address these hazards, the pharmaceutical industry has adopted a performance-based strategy using exposure control limits. This approach is based on establishing qualitative or semiquantitative criteria for assessing risk associated with the compounds and matching that information with known exposure-control approaches [Naumann et al. 1996].

Many of the processes used in pharmaceutical production are similar to those used in the nanoparticle industries discussed above and include blending, mixing, and handling of hazardous compounds in liquid and powder form. The general control concepts required for working with hazardous materials include specification of general ventilation, LEV, maintenance, cleaning and disposal, PPE, IH monitoring, and medical surveillance [Naumann et al. 1996]. Particular work practices, such as using HEPA-filtered vacuums instead of dry sweeping, are required. In addition, routine IH and medical monitoring ensure that work practices and engineering controls are effective.

Source containment is considered the highest level in the containment hierarchy and is used by the pharmaceutical industry [Brock 2009]. This category contains many options including elimination, substitution, product modifications, process modifications, and equipment modifications. These steps could include reworking the process to reduce the number of times material is transferred or keeping the product in solution to minimize aerosolization potential. The next level of control for capturing process emissions is the use of engineering controls such as glove boxes, downflow booths, and local exhaust ventilation.

Genaidy et al. [2009] conducted a detailed hazard analysis of a CNF manufacturing process and suggested the following potential sources of workplace exposure to nanomaterials:

- Leakage and spillage from reactors and powder processing equipment
- Manually harvesting product from reactors

- Discharging product into containers
- Transporting containers of intermediate products to the next process
- Charging the powders into processing equipment
- Weighing out powder for shipment
- Packaging material for shipment
- Storing material between operations
- Cleaning equipment to remove debris stuck to side walls
- Changing filters on dust collection systems and vacuum cleaners
- Further processing of products containing nanomaterials (e.g., cutting, grinding, drilling)

This detailed analysis, along with the review of exposure assessment studies in nanomaterial production and downstream user facilities described below, identify common processes that may lead to worker exposure to nanomaterials. This section provides some information on engineering control approaches that may be applicable for these common processes/tasks. Table 1 shows a generic process list along with applicable engineering controls and references. The engineering control column provides a framework for identifying exposure controls for particular processes. The third column shows the industry in which these control approaches have been tested. References are listed in the fourth column for studies that apply to each of these processes and controls.

Table 1. Engineering controls and associated tasks for various industries

Process/task	Engineering control	Industry	Reference
Reactor fugitive emissions	Enclosure	Nanotechnology	Tsai et al. 2009b Lee et al. 2011
Product harvesting	Glovebox	Nanotechnology	Yeganeh et al. 2008
Reactor cleaning	Spot LEV system/fume extractor	Nanotechnology	Methner 2008
Small-scale weighing	Chemical fume hood	Nanotechnology	Tsai et al. 2009a Ahn et al. 2008 Tsai et al. 2010
	Biological safety cabinet	Nanotechnology and laboratory	Cena and Peters 2011 Macher and First 1984
	Glovebox isolator	Pharmaceutical	Walker 2002 Hirst et al. 2002
	Nano fume hood	Pharmaceutical	
	Air curtain isolation hood	Nanotechnology/research	Tsai et al. 2010
Product discharge/bag filling	Discharge/collar hood	Silica and pharmaceutical	ACGIH 2013 HSE 2003e Hirst et al. 2002
	Continuous liner	Pharmaceutical	Hirst et al. 2002
	Inflatable seal	Pharmaceutical	Hirst et al. 2002
Bag/container emptying	Bag dump station	Silica	HSE 2003d Heitbrink and McKinnery 1986 Cecala et al. 1988
Large-scale weighing/handling	Ventilated booth	Pharmaceutical	Hirst et al. 2002 Floura and Kremer 2008 HSE 2003b
Nanocomposite machining	High velocity-low volume	Woodworking	
	Wet suppression	Nanotechnology	Bello et al. 2009
Air filter change-out	Bag in-bag out	Pharmaceutical	

3.3 Ventilation and General Considerations

It is important to confirm that the LEV system is operating as designed by regularly measuring exhaust airflows. A standard measurement - hood static pressure - provides important information on the hood performance, because any change in airflow results in a change in hood static pressure. For hoods designed to prevent exposures to hazardous airborne contaminants, the ACGIH *Industrial Ventilation: A Manual of Recommended Practice for Operation and Maintenance* recommends the installation of a fixed hood static pressure gauge [ACGIH 2010].

In addition to routinely monitoring the hood static pressure, additional system checks should be completed periodically to ensure adequate system performance, including smoke tube testing, hood slot/face velocity measurements, and duct velocity measurements using an anemometer. A dry ice test is another method of evaluation designed to qualitatively determine the containment performance of fume hoods. These system evaluation tasks should become part of a routine preventative maintenance schedule to check system performance. It is important to note that the collection and release of air contaminants may be regulated; companies should contact agencies responsible for local air pollution control to ensure compliance with emissions requirements when implementing new or revised engineering controls. To reduce the risk of exposure to nanomaterials, a few standard precautions should be followed in areas where exposures may occur:

- Isolate rooms where nanomaterials are handled from the rest of the plant with walls, doors, or other barriers.

- Maintain production areas where nanomaterials are being produced or handled under negative air pressure relative to the rest of the plant.

- Install hood static pressure gauges (manometers) near hoods to provide a way to verify proper hood performance.

- When possible, place hoods away from doors, windows, air supply registers, and aisles to reduce the impact of cross drafts.

- Provide supply air to production rooms to replace most of the exhausted air.

- Direct exhaust air discharge stacks away from air intakes, doors, and windows. Consider environmental conditions, especially prevailing winds.

3.4 Exposure Control Technologies for Common Processes

In a review of exposure assessments conducted at nanotechnology plants and laboratories, Brouwer [2010] determined that activities that resulted in exposures included harvesting (e.g., scraping materials out of reactors), bagging, packaging, and reactor cleaning. Downstream activities that may release nanomaterials include bag dumping, manual transfer between processes, mixing or compounding, powder sifting, and machining of parts that contain nanomaterials. Particle concentrations during production activities ranged from about 103–105 particles/cm^3. Most studies showed bimodal particle distributions with modes of about 200–400 nm and 1,000–20,000 nm, indicating that the emissions are dominated by aggregates and agglomerates. With the exception of leakage from reactors when primary manufactured nanoparticles may be released, workers are believed to be primarily exposed to agglomerates and aggregates.

Methner et al. [2010] summarized the findings of exposure assessments conducted in 12 facilities with a variety of operations: seven were R&D labs, one produced CNTs, one produced nanoscale TiO_2, one produced nanoscale metals and metal oxides, one produced silica-iron nanomaterials, and one manufactured nylon nanofibers. The most common processes observed at these facilities were weighing, mixing, collecting product, manual transfer of product, cleaning operations, drying, spraying, chopping, and sonicating.

Engineering controls used included portable vacuums with filters, laboratory fume hoods, portable LEV systems, ventilated walk-in enclosures, negative pressure rooms, and glove boxes. Tasks such as weighing, sonicating, and cleaning reactors showed evidence of nanomaterial emissions. The highest nanoparticle exposures measured occurred inside spray booth-type enclosures and during a spray dryer collection drum change-out. Other activities that resulted in higher exposures include reactor cleanout tasks (e.g., brushing and scraping slag material). Incidental (nonprocess) ultrafines were measured from a variety of sources, including electric arc welding, operating a propane-powered forklift, and the exhaust of a portable vacuum outfitted with filters.

From a review of published studies, some common sources of nanoparticles and fine particles can be identified. As expected, those processes that require material handling resulted in worker exposure to nanomaterials. Other activities that require operator interface with the reactor can result in nanoparticle exposure, and background concentrations may increase as a result of leakage from reactors under positive pressure. In addition, several studies found that evaluation of process emissions and exposure should take into account major sources of incidental nanoparticles that may be present in the workplace and also sources of natural nanoparticles, e.g., tree pollen brought into the work area through the facility HVAC system. Common incidental sources include diesel exhausts in outdoor air, welding fumes, forklifts, and gas-fired heaters. Several studies showed that the use of engineering controls can reduce operator exposure, while one study showed that a poorly designed enclosure actually increased exposure [Cena and Peters 2011; Methner et al. 2007; Tsai et al. 2009a, 2010; Yeganeh et al. 2008].

The following sections describe applicable engineering controls for common processes used by nanotechnology companies described in the literature. For each control, a background is given along with a summary of relevant research conducted on their performance. Many of the control concepts discussed in this section come from the HSE Control Guidance Sheets in COSHH Essentials: Easy Steps to Control Chemicals [HSE 2003a,b,c,d] and the ACGIH Industrial Ventilation Manual [ACGIH 2013]. Table 2 lists common processes and tasks, along with potential emission points and the section or figure(s) that address those processes.

Table 2. Process/tasks and emission

Process/task	Potential emission/ exposure points	See section	See figures
Production of bulk nanomaterials	Reactor fugitive emissions Product harvesting Reactor cleaning	3.4.1 3.4.1 3.4.1	7, 8 12
Downstream processing	Product discharge/bag filling Bag/container emptying Small-scale weighing Machining of products	3.4.3.1 3.4.3.2 3.4.2 3.4.3.4	14, 15, 16 17 10, 11, 12, 13
Product packaging	Small-scale weighing/handling Large-scale weighing/handling Product packaging	3.4.2 3.4.3.3 3.4.3	10, 11, 12, 13 18 14, 15, 16, 18
Maintenance	Facility equipment cleaning Air filter change-out Spill clean-up	3.4.4 3.4.4.1 3.4.4.2	 19

3.4.1 Reactor Operation and Cleanout Processes

Harvesting material from reactors has been identified as a potentially high exposure activity in several manufacturing plants [Demou et al. 2008; Lee et al. 2010, 2011; Methner 2008; Yeganeh et al. 2008]. In addition, cleanout of reactors has contributed to increasing facility concentrations and exposures to operation and maintenance workers. Leakage from pressurized reactors can also contribute to background concentrations and result in exposure to employees throughout the facility. When the reactors are small, some facilities have placed them inside fume hoods to help control fugitive emissions. Two studies have shown that when the reactor is housed in a well-designed and operated fume hood, particle loss to the work environment is low [Tsai et al. 2009b; Yeganeh et al. 2008]. When the reactors are larger, enclosures can be built that isolate the reactor from the environment and seek to reduce fugitive emissions (Figure 7).

Methner et al. [2010] summarized airborne measurements in 12 facilities that processed nanomaterials, including manufacturers and research and development labs. The authors found that some of the highest measured exposures occurred during reactor cleanout tasks, which included brushing and scraping slag material from the reactor walls and during torch cleaning. Demou et al. [2008] evaluated exposure to nanoparticles at a pilot-scale nanomaterial production facility. The major emission source was determined to be the production unit as the airborne particle concentrations rose when the unit was started and fell when production rate was decreased. The other task that resulted in substantial particle release was cleaning of the reactor using a vacuum cleaner not fitted with a HEPA filter. Evans et al. [2010] studied nanoparticle concentrations in a facility that manufactured and processed carbon nanofibers (CNFs). During the thermal treatment of the CNFs in a reactor under positive pressure, elevated concentrations of non-CNF ultrafines were released.

Figure 7. A large-scale ventilated reactor enclosure used to contain production furnaces to mitigate particle emissions in the workplace (Used with permission from Flow Sciences, Inc.)

Lee et al. [2010] conducted personal, area, and real-time sampling in seven CNT plants. Results showed that nanoparticles and fine particles were most frequently released upon opening the chemical vapor deposition (CVD) furnace. Catalyst preparation and the opening of the CVD furnace resulted in the release of nanoparticles in the range of 20–50 nm. Lee et al. [2011] also evaluated workplace exposures to nanoscale TiO_2 at manufacturing plants. In one TiO_2 plant, the reactor was small and was placed in a fume hood; the entire process was conducted in that hood. Even though the reactor was located in the hood, high concentrations of nanoparticles were measured outside the hood. Worker exposure increased during product harvesting because the worker put his head into the hood to brush out the product powder. A second TiO_2 plant isolated the large-scale reactor with a vinyl curtain and used a glove box for the harvesting of product from the reactor. Overall, airborne particle concentrations were fairly stable during production although increases occurred during both the operation of a process vacuum pump and welding conducted in the facility.

Yeganeh et al. [2008] evaluated a small facility producing carbonaceous nanomaterials including fullerenes. The process involved the production of materials in an arc furnace that was enclosed in a ventilated fume hood. This hood had a plastic front face shield and ports that allowed worker access during the process. The process involved placing graphite rods into the furnace, volatilizing the graphite in the furnace, producing raw soot, and using a scoop and brush to remove raw soot into a jar. At the beginning or end of each day, the reactors were completely cleaned by manual sweeping and vacuuming to remove residual soot. Real-time particle analyses showed that physical handling of material (sweeping of the reactor) resulted in the aerosolization of ultrafine particles. Measurements inside and outside the reactor enclosure (i.e., fume hood), however, showed that the hood was effective at containing particulates.

Methner [2008] evaluated the use of a portable LEV unit for controlling exposure during cleanout of a vapor deposition reactor used for producing nanoscale metal catalytic materials comprised of manganese, cobalt, or nickel. Following the automated collection of product materials, an operator cleaned out slag and waste product from the reactor using brushes and scrapers. Initial measurements had shown this task to be a high-exposure task for the operator. A follow-up survey was conducted at the facility using a commercially available fume extraction unit with HEPA filtration to pull airborne dusts away from the operator during cleanout. Analysis of real-time instrumentation and filter samples analyzed for metals showed an average reduction in airborne concentrations of 88%–96% during three cleanout procedures.

Emission sources related to reactor operations, harvesting, and maintenance can be categorized as fugitive or task–based. The approaches that have been used for controlling fugitive emissions from the reactor have primarily been ventilated enclosures. Laboratory fume hoods and glove boxes can be used when the reactor is small, typical of R&D or pilot operations. Where the production reactors are larger, custom-fabricated enclosures often constructed from a polycarbonate, transparent thermoplastic material, or vinyl curtains have been used to reduce emissions (Figure 7). When designing these types of enclosures, it is necessary to consider reactor access needs, determination of exhaust airflows capable of maintaining a negative pressure (even during the opening of access doors), and accommodation of heat loads generated by the process. Failure of containment can result from not carefully addressing these key design needs. When looking at pressure differentials, it is important to study the airflow to minimize turbulent situations that can actually increase particle release rather than containing the particles.

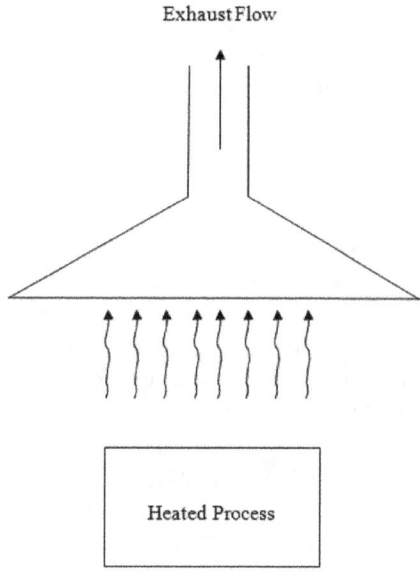

Figure 8. A canopy hood used to control emissions from hot processes

When a process is heated, the use of canopy hoods (Figure 8) may be another reasonable alternative as long as the design meets the operational and facility exposure control requirements [ACGIH 2013; McKernan and Ellenbecker 2007]. Even if the process does not involve heat, contaminant capture velocities suitable for gas/vapor contaminants (rather than coarse particulates) may be sufficient, as ultrafine and nanoparticles possess negligible inertia and follow the flow stream well.

When controlling exposures during operations such as product harvesting and reactor cleanout, solutions such as spot LEV systems (e.g., a fume extractor) or containment may be acceptable alternatives. Manual harvesting of product materials may be better suited for higher-level enclosure controls such as a glove box or a specially designed enclosure to provide good capture while minimizing loss of product materials. The use of a commercially available fume extractor has been shown to be effective in reactor cleanout and provides a flexible solution that may meet facility needs across a range of operations [Methner 2008]. Selection of any control should be evaluated to ensure worker acceptance and use as well as verifying that it meets the exposure control objectives.

3.4.2 Small-scale Weighing and Handling of Nanopowders

Small-scale weighing and handling of nanopowders are common tasks; examples include working with a quality assurance/control sample and processing small quantities in downstream industries. During these operations, workers may weigh out a specific amount of nanomaterials to be added to a process such as mixing or compounding. The tasks of weighing out nanomaterials can lead to worker exposure primarily through the scooping, pouring, and dumping of these materials. Many different types of commercially available laboratory fume hoods can be employed to reduce exposure during the handling of nanopowders. Other controls have also been used in the pharmaceutical and nanotechnology industries for containment of powders during small-quantity handling and manipulation. They include glove boxes, glove bags, biological safety cabinets or cytotoxic safety cabinets, and homemade ventilated enclosures.

Methner et al. [2007] evaluated a university-based research lab that used CNFs to produce high-performance polymer materials. Several processes were evaluated during the survey: chopping extruded materials containing CNFs, transferring and mixing CNFs with acetone, cutting composite materials, and manually sifting oven-dried CNFs on an open benchtop. Real-time monitoring did not identify any process as a substantial source of airborne CNF emissions; however, weighing/mixing of CNFs in an unventilated area resulted in elevated particle concentrations compared to background. Other studies have shown that benchtop activities such as probe sonication of nanomaterials in solution can also result in emission of airborne particles [Johnson et al. 2010; Lee et al. 2010]. Producing dispersions by sonication is a primary operational step, and the industrial hygiene assessment should address the sound level exposure as well as the potential exposure to aerosols of nanomaterials from the sonication. Maintaining the sonicator/dispersion process within an enclosure such as a hood can be an effective means for mitigating the noise and aerosol exposure.

3.4.2.1 Fume Hood Enclosures

In 2006, a survey was conducted of international nanotechnology firms and research laboratories that reported manufacturing, handling, researching, or using nanomaterials [Conti et al. 2008]. All organizations participating in the survey reported using some type of engineering control. The most common exposure control used was the traditional laboratory fume hood with two-thirds of firms reporting the use of a fume hood to reduce exposure to workers. These devices have been used for many years in research laboratories to protect workers from chemical and biological hazards. The design and operation of the fume hood is an important factor when considering good exposure control. Traditional designs for laboratory fume hoods create airflow patterns that form recirculation regions inside the hood. In addition, airflow around the worker, as shown in Figure 9, creates a negative pressure region downstream of the worker, which may provide a mechanism for the transport of materials out of the hood as well as into the breathing zone of the worker.

Recent research has shown that the laboratory fume hood may allow the release of nanomaterials during their handling and manipulation [Tsai et al. 2009a]. This research evaluated exposures related to the handling (i.e., scooping and pouring) of powder nanoalumina and nanosilver in a constant air volume (CAV) hood, a bypass hood, and a variable air volume (VAV) hood. This study showed that the CAV fume hood, in which face velocity varies inversely with sash height, allowed the release of significant amounts of nanoparticles during pouring and transferring activities involving nanoalumina. The particles that escaped the fume hood were circulated to the general room air and were not cleared by the general ventilation system for 1/2–2 hours. Sash heights both above and below the recommended height (corresponding to a face velocity of 80–120 ft/min) may lead to increased potential exposure for the user. In contrast, more modern hoods such as the VAV hood, which is designed to maintain the hood face velocity in a desired range regardless of sash height, yielded better containment of nanoparticles than the other hoods tested.

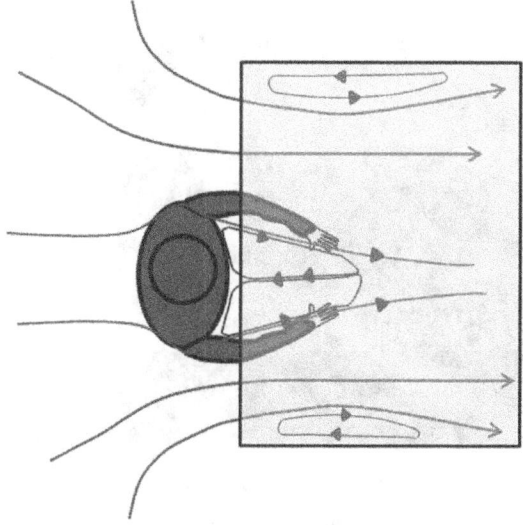

Figure 9. Schematic illustration of how wakes caused by the human body can cause transport of air contaminants into the worker's breathing zone

A meta-analysis of fume hood containment studies was conducted to identify the important factors that affect the performance of a laboratory fume hood [Ahn et al. 2008]. An analysis of factors affecting the containment performance of the hoods showed that worker exposures to air contaminants can be greatly impacted by a variety of operational issues. Increasing the distance between the contaminant source and the breathing zone leads to reduced exposure. Exposures can also be reduced by limiting the height/area of the sash opening; increasing the height of the sash opening increased the risk of hood containment failure. The presence of a manikin/human subject in front of the hood caused the greatest risk of hood failure among factors studied. This indicates that containment testing should include an operator or manikin to adequately assess hood performance. Face velocity did not make a significant difference in hood performance unless it was extremely high or low (> 150 ft/min or < 60 ft/min). Several hood operating factors showed an effect but were not statistically significant, including sash movement, hand and arm movement, pouring/weighing, and thermal load.

New fume hoods specifically designed for nanotechnology are being developed primarily based on low-turbulence balance enclosures, which were initially developed for the weighing of pharmaceutical powders. The use of bench-mounted weighing enclosures, as seen in Figure 10, is common for the manipulation of small amounts of material. These fume hood-like LEV devices typically operate at airflow rates lower than those in traditional fume hoods and use airfoils at enclosure sills to reduce turbulence and potential for leakage. They also have face velocity alarms to alert the user to potentially unsafe operating conditions. Based on the hazards assessment, these fume hood-like LEV devices can be outfitted with HEPA filtration or connected to the ventilation exhaust system.

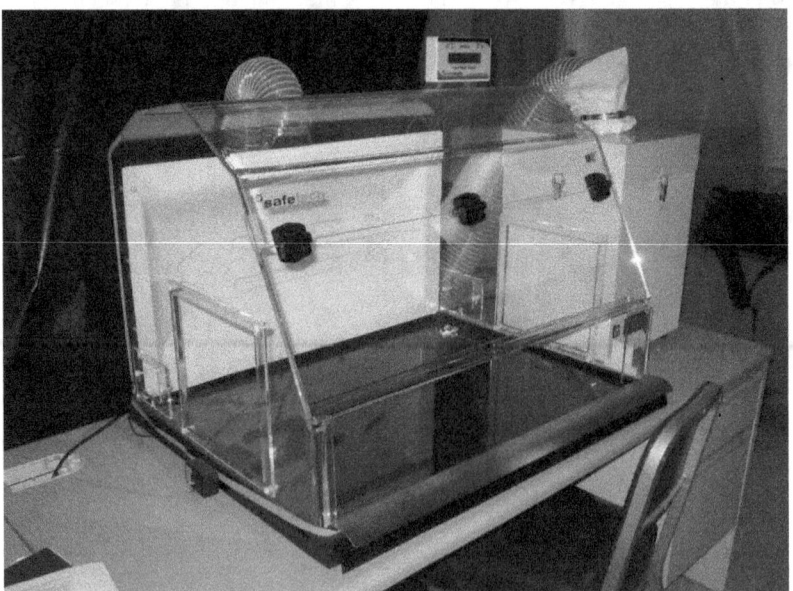

Figure 10. Nano containment hood adapted from a pharmaceutical balance enclosure

3.4.2.2 Biological Safety Cabinets

The Centers for Disease Control and Prevention (CDC) divides biological safety cabinets (BSCs) into three classes: Class I, Class II, and Class III. The Class II BSCs are further divided into four subcategories (A1, A2, B1, B2) [DHHS 2009]. These hoods are used for processes that require operator and product protection. The BSC pulls air into the hood to protect the operator while providing a downward flow of HEPA-filtered air inside the cabinet to minimize cross-contamination along the work surface (see Figure 11). The most common BSC (Type II/A2) uses a fan to provide a curtain of HEPA-filtered air over the work surface. The downward moving air curtain splits as it approaches the work surface; some of the air is drawn to the front exhaust grille and the remainder to the rear grille. The air is then drawn back up to the top of the cabinet where it is recirculated or exhausted from the cabinet. In general, 70% of the air is HEPA-filtered and recirculated while 30% is filtered and then exhausted from the cabinet. The make-up air is drawn through the front of the cabinet. The air being drawn in acts as a barrier to protect the workers from contaminated air leaking out of the hood.

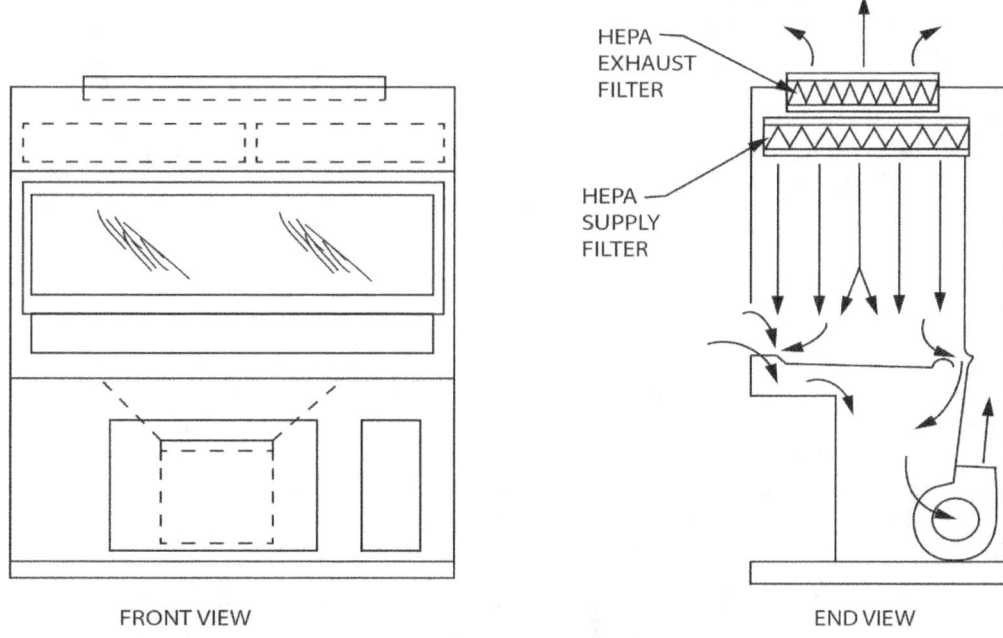

Figure 11. A tabletop model of a Class II, Type A2 biological safety cabinet (BSC) (Used with permission from ASHRAE [2011].)

Cena and Peters [2011] evaluated the effectiveness of ventilated enclosures including a Class II, Type A2 BSC and a custom fume hood during the manual sanding of epoxy test samples reinforced with CNTs. Sanding of CNT-epoxy materials released respirable-sized (micron-sized) particles but generally no nano-sized particles. The respirable mass concentration in the operator's breathing zone while using the BSC was approximately two orders of magnitude lower than the concentration when using the custom fume hood. The use of the custom fume hood resulted in an increase of breathing zone concentrations of about one order of magnitude compared to the use of no controls. The custom fume hood had a low average face velocity of about 45 ft/min with high variability across the hood face. The authors suggested that the poor performance of the custom fume hood may have been due to its rudimentary design, which did not include a front sash or rear baffles. The lack of these common fume hood features along with the low average face velocity may have resulted in poor airflow distribution across the face and increased leakage.

Macher and First [1984] evaluated the effect of airflow rates and operator activity on containment effectiveness for a Class II, Type B1 biological safety cabinet using bacterial spores released by two 6-jet collison nebulizers. The hood sash height correlated negatively with the containment effectiveness; that is, the higher of two sash heights provided better containment of the aerosol. In addition, working in the front half of the cabinet provided better protection than working in the rear half of the cabinet. The authors postulated that working in the rear of the hood caused the operator to move closer to the hood opening, blocking the opening and causing more turbulence and leakage around the sides of the hood. The operator withdrawing his arms from the hood caused significantly more leakage than moving arms side to side within the hood. The authors concluded that testing BSCs with persons working at them provides more information than static testing alone and that even well-designed cabinets lose a small fraction of aerosols.

3.4.2.3 Glove Box Isolators

A glove box isolator fully isolates (contains) a small-scale process and is sometimes referred to as a primary protection device (Figure 12) [HSE 2003a]. The design can be either the more typical hard unit or a soft, flexible containment unit (often referred to as a glove bag). Glove boxes provide a high degree of operator protection but at a cost of limited mobility and size of operation. In addition, cleaning the glove box may be difficult, and, to prevent exposures, operators should use caution when transferring materials and equipment into and out of the glove box. In general, glove boxes include a pass-through port, which allows the user to move equipment or supplies into and out of the enclosure.

The performance of a glove box containment system was evaluated during weighing activities of fine lactose powder (a common pharmaceutical surrogate test material). Air samples were collected at four locations: inside the glove box, in the pass-through, in front of the glove box, and at the exit of the recirculating HEPA filter [Walker 2002]. The results of sampling a 10-minute task showed the average concentration measured inside the glove box was 298 µg/m^3, the average concentration in the integral pass-through was 35 µg/m^3, and concentrations measured in the room, including downstream of the glove box exhaust, were below the analytical limit of detection of 1 µg/m^3. Sample swabs of interior surfaces showed dust contamination within both the main glove box and pass-through. These results indicated that, although internal surfaces were contaminated with the materials, no leakage from the glove box was detected.

Figure 12. A glove box isolator for handling substances that require a high level of containment (Contains public sector information published by the Health and Safety Executive and licensed under the Open Government License v1.0.)

3.4.2.4 Air Curtain Fume Hood

A recent fume hood design addresses the known issues surrounding the recirculating flow patterns both inside the fume hood and around the operator (Figure 9). The air curtain-isolated fume hood, as shown in Figure 13, uses a push-pull ventilation configuration created by a narrow planar jet from the sash to an exhaust slot along the base of the hood opening. Tsai et al. [2010] evaluated the performance of this hood during handling and manipulation of nanoparticles. In this test, measurements in the worker's breathing zone were taken while nanoalumina powders were manually transferred or poured between several 400-ml beakers. The air-curtain hood had very low particle release during all tested conditions (i.e., varying sash heights) with low but measurable release occurring at the lowest sash position. This same study showed that the particle leakage from two traditional fume hoods (both a CAV and VAV hood) exhibited substantial particle release during similar nanomaterial handling operations. This study suggested that the air curtain isolated hood may provide better containment performance during typical handling procedures.

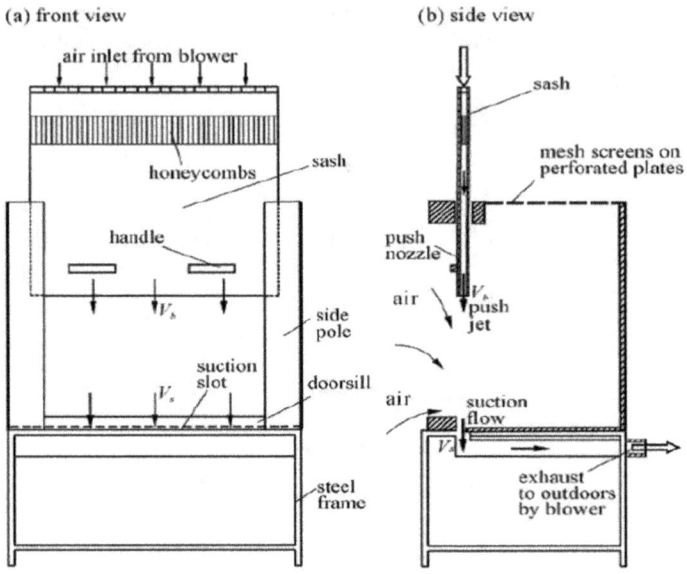

Figure 13. Air curtain safety cabinet hood that uses push-pull ventilation (Used with permission from Huang et al. [2007a].)

3.4.2.5 Summary

Overall, the published studies suggest that the selection of a fume hood with improved operating characteristics such as a VAV hood provides better operator protection than conventional fume hoods when handling dry nanomaterials. When using any hood, the worker should strive to maintain the face velocity in the recommended range of 80–120 ft/min [ACGIH 2013]. Additionally, proper use of the engineering control by the operator and validation of the performance of the control equipment are essential for risk mitigation. Newer nano hoods based on pharmaceutical weigh-out enclosures may be a reasonable alternative to larger fume hoods when only small-scale, benchtop manipulation of powders is needed. These hoods have proven effective in the pharmaceutical industry but need more thorough evaluation to assess the impact of lower airflow rates on containment performance, especially in the industrial environment. BSC-type hoods are commonly used for containing hazardous powders in hospitals for hazardous drug formulation and may have features designed to improve containment performance over traditional fume hoods. However, there are few published studies on their effectiveness for containing nanomaterials. The selection of a BSC appropriate for use with nanomaterials is essential. Considerations including how to clean the cabinet after use, how to maintain the BSC during required maintenance such as filter change-outs, and proper exhaust configuration (sending exhaust out of the production area versus recirculating exhaust air) should be considered prior to use. Glove box isolators typically provide a greater level of worker protection but at a cost of reduced access and limited operational scale. Newer hood designs, such as the air curtain fume hood, have shown excellent containment performance in initial studies and may be potential control options in the future.

Many options are available to facilities that require worker protection during small-scale material handling operations. The best option for a given process depends on several factors including scale of handling operations, physical properties of materials being handled (size, density, wet or dry formulation), work environment (lab versus plant, cross drafts, nearby activity), equipment requirements (size of equipment/operation being enclosed), and level of protection required. Independent of the control selected, users should also adopt good work practices, such as using the smallest possible quantities of materials. Other procedures, such as wiping down and sealing containers before they are removed from the enclosure, are recommended. In addition, using care when working with powders, such as refraining from dropping dust from height, helps to prevent dust generation and to reduce operator exposure. The proper positioning of these workstations away from doors, windows, air supply registers, and aisleways will help to reduce the impact of cross drafts.

3.4.3 Intermediate and Finishing Processes

Exposures resulting from the manual handling of powdered materials are common in industry. Reduction in worker exposure through implementation of careful work practices and appropriate engineering controls would benefit these operations. Dumping bags of powdered materials has been commonly reported in the literature for production and processing. Typically, a worker dumps the ingredients for one process into a hopper and then compacts or disposes of the empty bags. Ventilated bag-dumping stations have been used successfully in a variety of industries and applications. The transfer of large quantities of nanomaterials requires different solutions adapted to the particular process. However, a few controls that are applicable to these common processes are available and have been evaluated for similar industrial operations.

After the completion of production, many nanomaterials are sent for further processing. The powder product may be refined through a common process such as spray drying [Lindeløv and Wahlberg 2009]. Other studies have documented collection of fugitive emissions of nanomaterials from the process reactors using devices such as baghouse air filters [Evans et al. 2010]. In both of these operations, the nanomaterials are collected in a barrel or drum following the completion of these production steps. Several examples of engineered drum or bag filling solutions have been described elsewhere and could be implemented to reduce such releases [ACGIH 2013; Hirst et al. 2002]. These engineering controls consist of enclosing the product off-loading process by temporarily sealing the drum/bag to the filling vessel above and/or overbagging through a continuous liner type bagging system. The addition of a local exhaust ventilation hood near the drum/bag opening could also be used to capture airborne nanomaterials.

Evans et al. [2010] studied nanoparticle concentrations in a facility that manufactured and processed carbon nanofibers (CNFs). The authors discussed four discrete events that resulted in elevations in airborne particle concentrations. The largest increases in particle concentration measured within the plant were related to manual handling processes, such as dumping product into lined drums and manual change-out and closing bags of final treated CNF product. Increases in particle concentrations were the result of the change-out and closing of the collection bag containing approximately 15 lbs of CNFs. Emissions from this event were almost entirely due to aerosolized CNFs. Tamping of the bag to settle contents (so that it could be adequately closed) and subsequent closing appeared to efficiently aerosolize CNF material through the bag opening into the workplace environment. This resulted in an increase in respirable mass concentration and a dark visible airborne CNF plume.

A few studies have been conducted to look at the emission of nanoparticles from downstream products during machining of nanocomposites. Methner et al. [2007] reported increases in total carbon (a marker for nanoparticles), particle number, and mass concentration during the wet sawing of a CNF-impregnated composite. However, the increase in particle concentration was primarily of particles greater than 400 nm in diameter. Vorbau et. al. [2009] evaluated nanoparticle release from oak and steel panels coated with polyurethane mixed with zinc oxide nanoparticles. A standard abrasive test rig was used to provide uniform conditions for testing the release of particles from the surface of the panels. During the abrasion tests, no significant release of particles below 100 nm was observed. However, the nanoscale zinc oxide particles were embedded in the aerosols with larger surface area. Bello et al. [2009] evaluated the release of nanoscale particles during dry and wet cutting of nanocomposite materials. Two composites were used for evaluation: a CNT-enhanced graphite prepreg laminate sheet and a woven alumina fiber cloth with CNTs grown on the surface of the fibers. Significant exposures to nanoscale particles were generated during dry cutting of all composites with emission levels being related to composite material and thickness; wet cutting reduced exposures to background levels.

For all processes/tasks discussed, engineering controls should be adapted for the specific process. Acceptable exhaust volumes and capture velocities may differ from currently available guidance due to differences in materials being handled. Pilot testing of any controls should be conducted to evaluate proper control operation and verify that exposures are controlled to desired levels.

3.4.3.1 Product Discharge/Bag Filling

The process of filling bags with nanomaterials is commonly done following large-scale production or refining processes. The off-loading of product after spray drying, for example, may be a significant source of exposure when post-processing nanomaterials. In the spray-drying process, a mixture of liquid and powder ingredients (slurry) is sprayed within a large sealed tank. Heat within the tank dries the slurry droplets, leaving a powder as the finished product. When the process is completed, the powder product is commonly discharged into a bulk tote or drum before packaging. Methner et al. [2010] reported exposure measurements at 12 facilities and noted that the highest background-adjusted concentration was observed during spray dryer drum changeout. Evans et al. [2010] reported exposures related to changing out a drum that collected fugitive CNF materials from a process reactor using a baghouse filtration system. Even though the processes differ, the tasks for each of these steps are similar and include the removal of the drum from the process outlet. These drums are often sealed to the process outlet minimizing exposure during production but potentially expose workers when removing the drums or liners.

A ventilated, collar-type hood around the discharge point can help minimize worker exposure to dust. Figure 14 presents a control approach for filling bags with solid powder materials [HSE 2003c]. The control includes the specification of a ventilated enclosure around the powder discharge outlet and applies to filling smaller product bags as well as intermediate bulk containers. This design guidance recommends an inward air velocity of 200 ft/min (1.0 m/s) into the enclosure. The ACGIH *Industrial Ventilation Manual* [ACGIH 2013], Design plate VS-15-02, Bag Filling, is similar in design to the HSE exhaust hood (Figure 14) but specifies an overall hood flow rate of 400–500 ft³/min for nontoxic dust or 1,000–1,500 ft³/min for toxic dust with a maximum inward air velocity of 500 ft/min. These flow rates have been specified for common industrial powders and may need to be adjusted based on the process and properties of the nanoscale materials being addressed to prevent excessive loss of product.

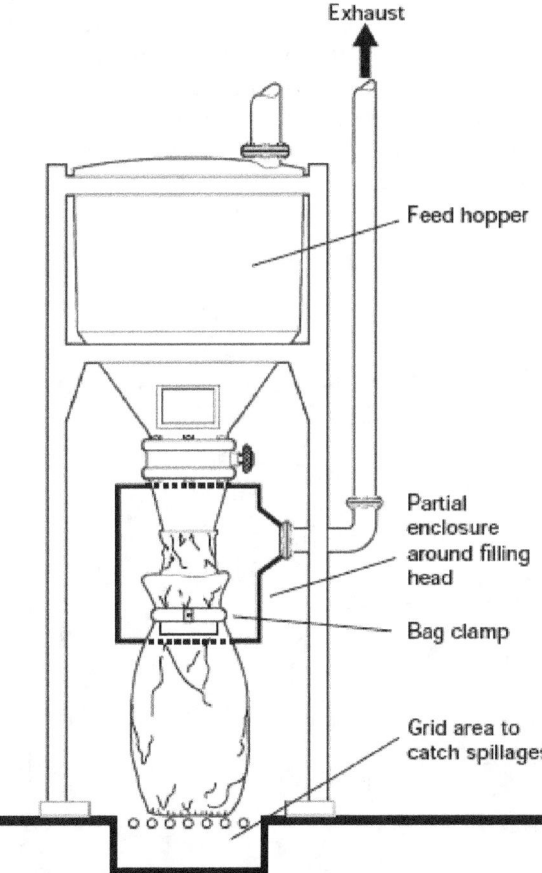

Figure 14. Ventilated collar-type exhaust hoods for containing dust during product discharge or manual bag filling (Contains public sector information published by the Health and Safety Executive and licensed under the Open Government License v1.0.)

In addition to ventilation solutions, other dust control approaches have been used in a variety of industries and should be applicable for nanomaterial production. For example, an inflatable seal can be used to create a dust-tight seal on the discharge outlet of a spray dryer during the product discharge/bag filling process (Figure 15). The seal inflates during the product transfer from the process to the packaging bag (providing the seal) and deflates once the transfer is completed to allow removal of the bags. These systems are available on many commercially available bulk bag filling systems [Hirst et al. 2002].

Another system that can be used to contain powders during process off-loading/emptying is the continuous liner system (Figure 16). Polypropylene liners are often used when products are discharged from the industrial processes into the intermediate or final product containers. In this operation, a sleeve of polypropylene liners is stowed around the circumference of the discharge outlet. The first liner, the bottom having been sealed, is pulled down into the overpack (usually a drum or a cardboard box). Product is discharged into the liner through a butterfly valve on the process outlet. Once full, the top of the first liner sleeve is closed using tape or a fastener, or it is heat sealed and cut. The product is sealed within the poly-lined container, and a new sealed poly liner is pulled down to start discharge into the next container. This continuous process seals off the primary leak paths for dust during unloading of an industrial blender or other equipment. These systems are commonly used in the pharmaceutical industry and may provide cost-effective alternatives to traditional local exhaust ventilation control systems for nanotechnology facilities.

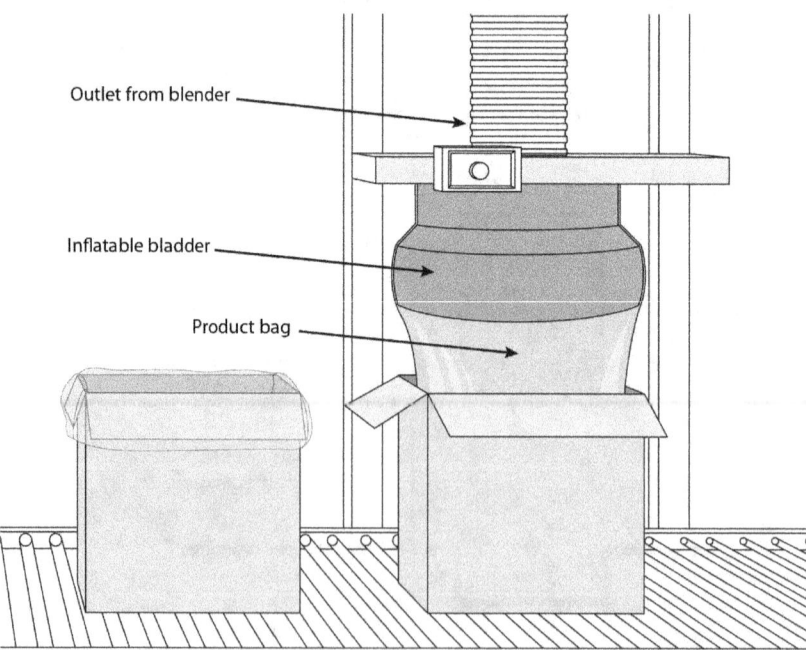

Figure 15. An inflatable seal used to contain nanopowders/dusts as they are discharged from a process such as spray drying

Figure 16. A continuous liner product off-loading system that uses a continuous feed of bag liners fitted to the process outlet to isolate and contain process emissions and product (Used with permission from ILC Dover.)

3.4.3.2 Bag Dumping/Emptying

Technology used to control dusts during bag dumping has been in place for many years. The standard control—a ventilated bag dump station—consists of a hopper outfitted with an exhaust ventilation system to pull dusts away from workers as they open and dump bags of powdered materials.

This equipment eradicates the dust problems caused by manually emptying bags and the need to dispose of empty bags. This ensures a healthy environment is maintained in the process area as well as reduces maintenance and repair problems caused by powder contamination to surrounding areas. The basic equipment consists of a bag dump cabinet with a dust extraction outlet for connection to a separate dust collector or existing plant exhaust. A bag is placed on the mesh support shelf and manually slit, with the contents falling directly into the inlet of a flexible conveyor or mixing tank. A side-mounted empty bag compactor may also be included. Design examples for these devices are available from several manufacturers of industrial materials. The British HSE has developed a control approach for a ventilated station for emptying bags of solid materials [HSE 2003d]. The control includes the specification of a face velocity of 200 fpm (1.0 m/s) and includes a waste bag collection chute (Figure 17).

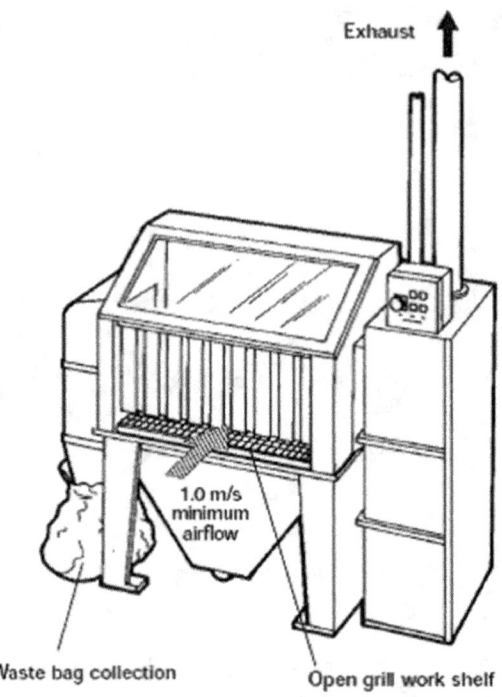

Figure 17. A ventilated bag-dumping station that reduces dust emissions when emptying product from bags into a process hopper (Contains public sector information published by the Health and Safety Executive and licensed under the Open Government License v1.0.)

Research into the effectiveness of these types of devices has shown that worker exposure to dust and vapors can be reduced. A review of commercially available units showed that their use with a variety of materials—including limestone, carbon black, and asbestos—controlled particle concentrations to acceptable levels [Heitbrink and McKinnery 1986]. However, particle contamination on the surface of the bag and handling/disposal of bags caused increased worker exposure. An integral pass through to a bag disposal chute/compactor can help reduce dust exposure resulting from bag handling. Further studies in mineral processing plants showed that the use of an overhead air supply also significantly decreased worker exposure [Cecala et al. 1988]. The ACGIH *Industrial Ventilation Manual* also has two designs that are applicable to the control of powder materials during bag dumping [ACGIH 2013]. Design plate VS-15-20, Toxic Material Bag Opening, is similar in design to the HSE station described above but recommends a slightly higher control velocity of 250 fpm at the face of the station opening. In addition, Design plate VS-50-10, Bin and Hopper Ventilation, requires a hood face velocity of 150 fpm. In general, higher velocities are specified to adequately capture dusts in a plant environment. While the materials used in the studies discussed above were not nanoscale, the application of the dust control concept is still relevant. However, the capture velocities specified in the *Industrial Ventilation Manual* may be excessive when attempting to contain nanomaterials; lower velocities may be warranted.

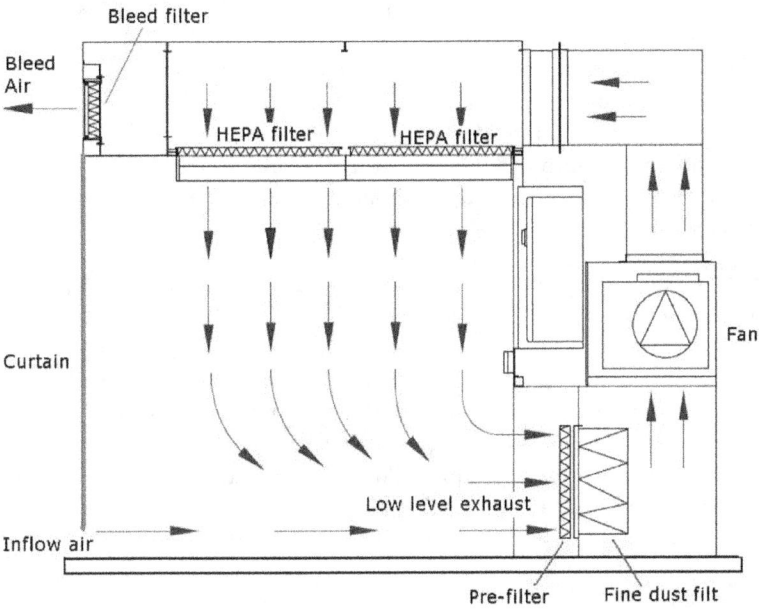

Figure 18. A unidirectional downflow booth for handling larger quantities of powders (Used with permission from Esco Technologies, Inc. [2012].)

3.4.3.3 Large-scale Material Handling/Packaging

Unidirectional flow booths, or downflow booths, as seen in Figure 18, are used in pharmaceutical applications for large-scale powder packing, process loading, and tray dryer loading [Hirst et al. 2002]. Similar applications have been proposed for handling hazardous dye powders. In general, these booths supply air from overhead (commonly at 100 fpm) over the full depth of the booth. Particles generated by processes carried out in the booths are captured and carried to the exhaust registers, which are located along the back wall of the booth. For the nanotechnology industry, these booths may provide a flexible solution for several common processes, including packaging of materials, transferring materials between process containers, or loading materials into containers for post processing.

Floura and Kremer [2008] evaluated a downflow booth used for transferring 25 kg of lactose (a surrogate pharmaceutical material) from drum to drum inside a downdraft booth. Air samples were collected in the operator's breathing zone and around the perimeter of the process during the transfer operation. The operator scooped the lactose powder from the initial drum into the final product drum until it was nearly empty and then carefully inverted the bag to pour the remainder of the contents into the final container. With no active ventilation controls, the concentration within the operator breathing zone averaged 2,250 µg/m³. When the ventilation inside the booth was turned on, the breathing zone concentration was substantially reduced to an average of 1.01 µg/m³. Finally, the authors evaluated the downflow booth with a ventilated collar added. The ventilated collar surrounded the interface between the drums and exhausted air at a rate of 425 ft3/min. During this test, the initial drum was inverted and the powder materials were emptied by gravity with the operator massaging the materials into the final product drum. The operator's breathing zone concentration averaged 0.03 µg/m^3 during this process. This study showed that the use of a downflow booth significantly reduced operator exposure during powder transfer processes and that adding a second level of LEV, the ventilated collar, further reduced the exposure by two orders of magnitude.

HSE Control Guidance Sheet 202, Laminar Flow Booth, presents a design for powder-handling processes called a horizontal- or cross-flow design [HSE 2003b]. The concept behind the design is similar to the downflow booth except that air enters the booth from the booth face. Air moves across the back of the worker toward the back of the booth. An issue with the cross-flow design is the secondary airflow patterns caused by the presence of the operator in the booth. Additionally, if purity or cleanliness of the product is important, sweeping of the air across the operator could be problematic. These patterns may cause turbulent dispersion of dust in the booth and result in higher operator exposure or potential leakage, compared to the downflow booth, but may provide a reasonable control for some processes.

3.4.3.4 Nanocomposite Machining

Initial studies have shown that machining some nanocomposite materials can result in the release of nanoscale particles to the work environment. Engineering controls when machining materials are available for most common processes. They range from ventilation of handheld tools using a high velocity-low volume (HVLV) system to the use of wet cutting techniques commonly adopted for silica control during construction activities. The use of standard dust controls such as those described by the HSE for woodworking as well as those identified in the ACGIH *Industrial Ventilation Manual* for machining processes provide a source of guidance that can be used to identify controls for machining processes. Bello et al. [2009] showed that the use of wet suppression techniques during sawing of nanocomposites reduced exposures down to background levels.

3.4.3.5 Summary

Processing nanomaterials involves a variety of steps. Following the production process, bulk unrefined materials may be packaged and shipped for use or may be subject to further processing. These processes require handling and manipulation of nanomaterials and have been shown to be a point of exposure for workers. These processes typically are composed of a limited number of tasks that may result in exposure of workers to nanoparticles or their agglomerates.

Product discharge. When processes empty into a large container, there is a potential for exposure especially when removing the full drum. Several engineering controls are available for this process/task. Nonventilation controls, such as inflatable seals and continuous liner systems, reduce the possibility of exposure. Ventilation-based options include the ventilated collar or enclosure around the discharge point. These solutions have been used and evaluated in a variety of industrial settings and have been shown to effectively capture dusts when properly designed and implemented in the process.

Bag dumping/emptying. When raw, bulk nanomaterials receive further processing/refining, those materials are often dumped from containers such as drums or bags into hoppers that feed the downstream processing equipment. Ventilated bag dump stations have been in use in industry for many years and have been proven to be effective in controlling dusts. Several commercial vendors and sources of design guidance exist for these devices.

Large-scale handling/packaging. When nanomaterials are handled in quantities larger than those that can easily fit in a fume-type hood, a unidirectional flow booth can provide a suitable control to reduce worker exposure and mitigate a potential emission source. These booths are commonly used in the pharmaceutical industry and have also been employed for handling hazardous dye powders in industrial settings. They provide the flexibility for a variety of operations that require handling of nanomaterials from larger containers, such as drums. They can also be designed to provide local exhaust for specific operations that may occur inside the booths. These booths are available for a variety of commercial vendors or can be designed from sources of readily available guidance.

Machining of nanocomposites. When machining composite materials coated or impregnated with nanomaterials, good dust suppression techniques should be used. Guidance on dust suppression techniques from ventilation-based (woodworking-type) or mist/water-based (silica/construction-type) controls may be adopted to reduce worker exposures. Exposures during machining should be carefully monitored and controlled. Standard engineering controls may need modification to properly control emissions. In addition to engineering controls, workers may need to wear appropriate respiratory protection.

3.4.4 Maintenance Tasks

Maintenance of the production facility and equipment can lead to exposures that are often overlooked. Demou et al. [2008] noted that maintenance procedures were a source of considerable particle emissions, specifically during the vacuuming of a reactor using a vacuum cleaner with a high-efficiency filter. However, other researchers have observed that cleaning the process area after CNT preparation reduced airborne particle concentrations [Lee et al. 2010]. Another typical activity not reported in the literature is the changeout of facility air filters. When local exhaust ventilation is used to contain nanomaterials and dusts, facilities will typically use air filtration prior to exhausting air from the building or recirculating into the work zone. When filters require change-out, the use of integral containment equipment and procedures can reduce maintenance worker exposure. Other general maintenance procedures, such as modifying ductwork or performing fan maintenance, will also require appropriate precautions to avoid exposing workers to nanomaterials settled in the equipment. In addition, general good housekeeping processes and written spill response procedures can help reduce the potential for worker exposure.

3.4.4.1 Filter Change–out–Bag In/Bag Out Systems

Bag in/bag out procedures are typically designed to protect workers performing maintenance on air filter change out. Bag in/bag out housings are specifically designed to allow for removal of a dirty air filter while minimizing worker exposure [Filtration Group Inc. 2012]. In these systems, a plastic liner is attached to a service port on the filter unit, as shown on the following page in Figure 19. When the filter is ready for replacement, the facility maintenance worker, wearing appropriate PPE, removes the filter into a liner. This process contains the filter with its contaminants so the worker is not exposed and the particulates are not resuspended in the workplace environment.

3.4.4.2 Spill Cleanup Procedures

An organized, clean workplace enables faster and easier production, improves quality control, and reduces the potential for exposure. It is important to maintain good general housekeeping practices so that leaks, spills, and other process integrity problems are readily detected and corrected. Proper practices regarding spills include:

- Allowing only individuals wearing appropriate protective clothing and equipment and who are properly trained, equipped, and authorized for response to enter the affected area until the cleanup has been completed and the area properly ventilated.
- Using HEPA-filtered vacuums, wet sweeping, or a properly enclosed wet vacuum system for cleaning up dust that contains nanomaterials.
- Cleaning work areas regularly with HEPA-filtered vacuums or with wet sweeping methods to minimize the accumulation of dust.
- Cleaning up spills promptly.
- Limiting accumulations of liquid or solid materials on work surfaces, walls, and floors, to reduce contamination of products and the work environment.

Figure 19. Removal of a dirty air filter from a ventilation unit into a plastic bag to minimize worker exposure to particles captured by the filter unit (Used with permission from Filtration Group Inc. [2012])

CHAPTER 4
Control Evaluations

The effectiveness of engineering controls for reducing exposures to nanomaterials during manufacturing and handling has not been widely investigated. To evaluate control measures in nanomanufacturing facilities, investigators need to collect both quantitative and qualitative data to describe nanoparticle emissions. Accurate direct-reading instruments allow investigators to identify the source of contamination in real time for various task scenarios. More detailed information about the materials, such as morphology and chemical characteristics, can be obtained by collecting air filter samples for off-line analysis.

4.1 Approaches to Evaluation

Strategies for measuring nanomaterial exposures and emissions in the workplace are being developed and evaluated by a range of researchers [Brouwer et al. 2009; NIOSH 2009a; OECD 2009; Ramachandran et al. 2011]. Because there are currently no exposure limits for engineered nanomaterials in the United States, a multifaceted approach combining qualitative analysis with quantitative means should be used to determine nanoparticle emissions and control effectiveness [Oberdörster et al. 2005]. However, some researchers have suggested using non-mass-based metrics such as surface area or particle number as a reasonable approach to assessing health effects [Wittmaach 2007; Rushton et al. 2010; Oberdörster et al. 2005]. The evaluation procedures include (1) identification of emission sources, (2) background and area monitoring, (3) air concentration measurement by direct-reading instruments and filter-based sampling, and (4) measurement of air velocity and patterns.

4.1.1 Identification of Emission Sources

The main purpose of the initial assessment or walk-through survey is to identify potential sources of emissions and to help researchers prepare a sampling plan for the in-depth evaluation of processes and control measures. Portable direct-reading devices (e.g., hand-held CPCs and photometers) are recommended for quick identification. The initial assessment should involve looking at the processes and equipment as well as the general plant environment. To optimize and improve engineering controls, a control checklist is recommended for collecting basic information on methods, manufacturing processes, and existing controls.

4.1.2 Background and Area Monitoring

A plan to assess control effectiveness requires that measurements are first taken of background concentrations in adjacent work areas. This allows the contribution of individual processes to be assessed by removing the background component [Brouwer et al. 2004; Demou et al. 2008; Peters et al. 2009]. The background measurement should be repeated after process or task evaluations. High background concentrations need to be addressed before control evaluation. The following factors can affect background data:

- Monitoring period. In a nanomanufacturing facility, the day of the week or the time of day during the monitoring period will affect background levels since worker movement and frequency of worker operations are variable.

- Other activities or operations around the monitored activity. Any operation, such as product harvesting or equipment maintenance, in areas outside the monitoring location could potentially influence background concentrations at the monitoring location.

- General ventilation conditions. The layout and operation of the general ventilation system in the workplace should be considered while monitoring background concentrations. Basic ventilation data (e.g., volume of air flow, location of supply and exhaust, general air movement in the facility), including air supply source, should be collected. Additionally, variations in environmental conditions (especially humidity) need to be measured.

- Other sources. Some equipment can produce incidental (nonprocess) nanoparticles. Examples include diesel engines, welders, gas-fired heaters, and air compressors for pulse-jet baghouses or dust collectors.

Area (or static) monitoring can also be conducted to evaluate the general air quality of workplaces. Instruments such as the CPC and impactors are suitable for this type of monitoring. Ideally, filter samples can be taken at the same location as area monitoring with direct-reading instruments to make a side-by-side comparison.

4.1.3 Air Monitoring and Filter Sampling

The selection of direct-reading instruments (Table 3) for field evaluation must cover a wide range of particle sizes. Particle diffusion occurs rapidly when nanoparticles are released in the workplace. It results in nanomaterial agglomerates because of particle collisions. For example, the average particle size (or size distribution) is larger during product transfer than right after product harvesting. Based on the data collected during initial assessment, characterization of nanomaterial emissions can be conducted with direct-reading instruments to provide higher resolutions of spatial and time variation. To evaluate control efficiency for specific processes or tasks, the sampling ports should be located as close as possible to the suspected emission sources but outside of control measures (or at a worker's breathing zone). Filter sampling for off-line qualitative analysis must occur in parallel with real-time monitoring. Sampling duration may not be an issue for most direct-reading instruments but should be considered for filter sampling to avoid overloading. The data collected from an initial assessment can be used to determine sampling time and flow rate for filter samples.

Table 3. Summary of instruments and techniques for monitoring nanoparticle emissions in nanomanufacturing workplaces

Metric	Instrument	Remarks
Aerosol concentration	CPC	Real-time measurement. Typical concentration range of up to 400,000 particles/cm^3 for stand-alone models with coincidence correction; 100,000 particles/cm^3 for hand-held models.
	DMPS	SMPS often uses a radioactive source. FMPS uses electrometer-based sensors. Concentration range from 100–10^7 particles/cm^3 at 5.6 nm and 1–105 particles/cm^3 at 560 nm.
Surface area	Diffusion charger	Need appropriate inlet pre-separator for nanoparticle measurement. Total active surface area concentration up to 1,000 µm^2/cm^3.
	ELPI	Real-time size-selective detection of active surface area concentration. 2×10^4–6.9×10^7 particles/cm^3 depending on size range/stage.
Mass	Size selective static sampler	Low pressure cascade impactors. Micro-orifice impactors.
	TEOM	EPA standard reference equivalent method.
Aerosol concentration by calculation	ELPI	
Surface area by calculation	DMPS	
	DMPS and ELPI used in parallel	Surface area is estimated by difference in measured aerodynamic and mobility diameters.
Mass by calculation	ELPI	Calculated by assumed or known particle charge and density.
	DMPS	Calculated by assumed or known particle charge and density.

Abbreviations: CPC=condensation particle counter; DMPS=differential mobility particle sizer; SMPS=scanning mobility particle sizer; FMPS=fast mobility particle sizer; ELPI= electric low pressure impactor; TEOM=tapered element oscillating microbalance

4.1.4 Assessment of Air Velocities and Patterns

The measurement of air velocity and pattern is important to establish sampling locations, evaluate outdoor contaminant penetration, and assess the performance of existing control measures. Two widely used air fluid velocity measuring devices are the Pitot tube and the hot-wire anemometer. The Pitot tube is useful to measure flow in ducts with high temperatures and/or high particle concentrations, which could damage the thermal anemometer probe. Shown in Figure 20, a Pitot tube is a primary standard that measures total and static pressures, and air velocity is calculated by using the pressure difference (i.e., velocity pressure) based on the Bernoulli equation. The method for conducting a Pitot traverse is described in the ACGIH *Industrial Ventilation Manual* [ACGIH 2013]. The Pitot traverse is typically used to measure duct air velocity to estimate overall system exhaust flow rate. Occasionally it is difficult to find a suitable location for Pitot tube traverses. Accurate duct velocity can be obtained using this method; however, poor measuring locations will cause inaccurate estimates of exhaust air flow. Sometimes the airflow through a hood can only be determined by measuring the air velocity at the hood face.

The measurement of fume hood face velocity is an important method to assess proper operation and containment. The average hood face velocity can be measured by dividing the opening of the hood into equal area grids of approximately one square foot and logging the velocity at the center of each grid with the thermal anemometer. To measure the velocities at each grid point, the anemometer should be held perpendicular to the direction of air flow. An average face velocity can be calculated while the variation in hood velocity from grid to grid should be assessed and noted [ACGIH 2013; ASHRAE 1995].

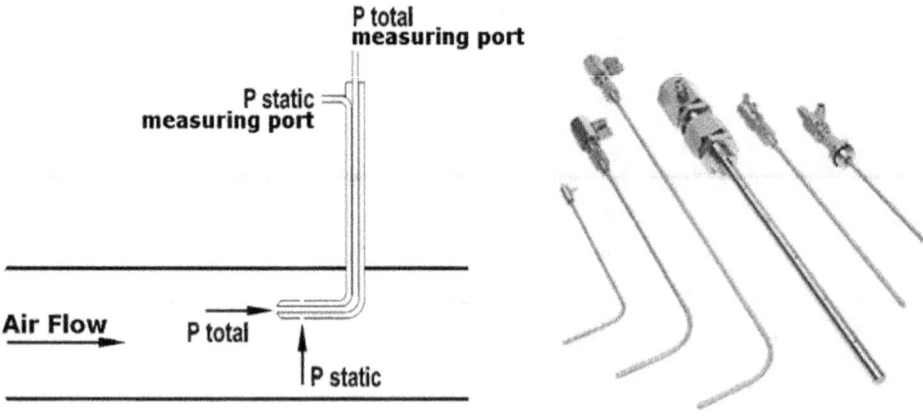

Figure 20. Operating principle of a Pitot tube (left) and different types of Pitot tubes (right)

In addition to Pitot tubes and anemometers for measuring air velocity, smoke generators provide a low-cost method to visualize airflow patterns around control measures. Figure 21 shows an example of a smoke generator. Airflow visualization techniques can be used to help understand the patterns of airflow in and around exhaust hoods and pressure differences between adjacent areas/rooms. Smoke can be released around the edge of, or inside of, a local exhaust hood to visualize the airflow patterns. This will help determine whether airborne particles are being effectively captured and removed by the ventilation system. Recorded observations should concentrate on (1) how much of the smoke is entrained into the LEV, (2) how quickly the exhaust captures the smoke, (3) the direction of air flow, and (4) whether or not any of the smoke visibly enters the worker's breathing zone. In addition, multiple replications of smoke-release observations should be made at locations where LEV performance is marginal or poor as indicated by reverse airflow, lack of air movement, slow clearance time, and escape of smoke from the hood. Special attention should be paid in subsequent tracer gas testing and air velocity measurements to locations where smoke release observations indicate poor or marginal capture efficiency. In addition, video may be taken of airflow visualization tests to provide feedback information to the company on system performance and factors that negatively affect hood performance.

Another use for airflow visualization is the evaluation of room pressurization status. It is recommended that rooms where nanomaterials are used be kept at an atmospheric pressure that is lower than adjacent areas. This condition helps contain the materials and reduce exposures to workers in other areas of the plant. Smoke should be released at the interfaces (doors or other openings) between any nanomaterial production areas and attached spaces. By releasing smoke at these interfaces, it can be easily observed whether air is moving into or out of the production area and proper remediation approaches may be implemented where necessary.

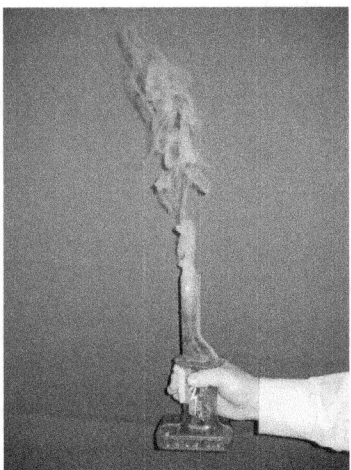

Photo by NIOSH   Figure 21. Smoke generator to visualize airflow

To qualitatively assess whether exhaust re-entrainment may be an issue, smoke can be released within each hood in the production room while a researcher observes the emission of the smoke through the exhaust stack. This qualitative test will help to evaluate the potential for re-entrainment of exhaust into any air intakes or roof openings. The behavior of the exhaust plume is dependent on varying environmental conditions such as wind speed and direction; therefore, this test should be repeated to capture the potential for re-entrainment under a variety of conditions. In addition, air velocity measurements should be taken at the center of the exhaust duct opening to evaluate the discharge velocity of the hood exhaust. These readings should be evaluated, along with the physical design and installation of the exhaust stack, against guidance from consensus standards organizations such as ASHRAE, ACGIH, or AIHA.

4.1.5 Facility Sampling and Evaluation Checklist

When evaluating a facility that manufactures or uses nanomaterials, it is important to first assess what engineering controls are in place in the facility. The initial assessment should involve looking at the processes and equipment as well as the general plant environment, the effective use of the engineering control by the operator(s), and the overall performance of the control equipment. Checklists are useful tools for helping to identify the process and facility factors related to nanomaterial production, use, emissions, and exposure. A checklist as shown in Table 4 may help for collecting basic process information (e.g., capacity, location, and usage) and control operation and maintenance parameters to ensure effectiveness of exposure control.

Table 4. Checklist of controls for nanomaterial manufacturing and handling

Item	Category	Data
Process/task	(select all applicable) ☐ Weighing ☐ Mixing ☐ Transferring ☐ Drying ☐ Cleaning ☐ Cutting/sanding ☐ Harvesting ☐ Unpacking ENMs ☐ Maintenance/repair ☐ Finishing (drilling, sawing, grinding) ☐ Packaging/shipping ☐ Others: _____	Workspace
		Duration (min)
		Frequency (times per day)
		Number of workers involved
		PPE type
	Background	Concentration Size distribution: _____ Number: _____ Mass: _____
Nanomaterial	☐ SWCNT ☐ MWCNT ☐ Other carbon-based ☐ Metals ☐ Oxides ☐ Quantum dots ☐ Composite: _____ ☐ Others: _____	Processing rate/volume
		Primary particle size
		Concentration at source Number: _____ Mass: _____
		Concentration at worker breathing zone or area (designate) Number: _____ Mass: _____ Breathing zone: _____ Area: _____

(Continued)

Table 4 (Continued). Checklist of controls for nanomaterial manufacturing and handling

Item	Category	Data
Control type	☐ NONE ☐ Local exhaust ☐ General exhaust/dilution ☐ Ventilated enclosure ☐ Fume hood ☐ Dust collector ☐ Laminar room ☐ Glove box ☐ Booth ☐ Other:_____	Dimensions Location Operation ☐ Hood type ☐ Face velocity: _____ ☐ Flow rate: _____ ☐ Temp: _____ ☐ Enclosure integrity ☐ Airflow patterns ☐ Recirculation Fan/filtration information Filter type: _____ Manufacturer: _____ Resistance (pressure drop): _____ Nominal design flow rate: _____ Fan type: _____ Flow rate: _____ Stack position/design:
Visual observation	Workspace Surface contamination Housekeeping Layout	

(Continued)

Table 4 (Continued). Checklist of controls for nanomaterial manufacturing and handling

Industrial Exhaust Ventilation Deficiency Report Worksheet	
☐ Building: ☐ Room: ☐ Hood number:	
☐ Date: ☐ Investigator/reporter: ☐ Fan number:	
Notes and sketch	

Management	Ductwork
☐ No local cognizant person ☐ Lack of records ☐ Lack of up-to-date plans and specifications ☐ Lack of emergency plan ☐ Insufficient employee training ☐ No hood testing mechanism ☐ No hood-use approval mechanism	☐ Holes, air leaking ☐ Dents ☐ Poor construction ☐ Plugged ☐ Corroded ☐ Leaking ☐ Dampers improperly set ☐ Fire dampers ☐ Doesn't meet SMACNA qualifications
Hood	**Fan/Motor**
☐ Improper type for operation/chemicals used ☐ Air leaking from hood (smoke noncontainment) ☐ Surfaces corroded ☐ Surfaces dirty ☐ Hood mechanisms inoperable ☐ Lack of real-time airflow monitor ☐ Flammable construction materials ☐ Slots not open to appropriate size ☐ Slots blocked by equipment, chemicals	☐ Worn out or corroded ☐ Insufficient rpm ☐ Belts slipping or broken ☐ Motor burned out ☐ Undersized fan
Hood operations	**Stack**
☐ Use of hood when hood exhaust off ☐ Hood not being used ☐ Inappropriate materials/equipment in hood ☐ Noisy	☐ Not attached ☐ Inappropriate location ☐ Inadequate height ☐ Stack exit velocity insufficient ☐ Aesthetic enclosure hinders dispersion
Work practices	**Exhaust hood**
☐ Untrained personnel ☐ Rapid movements at hood face ☐ Placing upper body in hood ☐ Operating outside hood	☐ Inadequate exhaust volume ☐ Inadequate face velocity ☐ Inadequate face velocity range ☐ Turbulence in hood face

(Continued)

Table 4 (Continued). Checklist of controls for nanomaterial manufacturing and handling

Industrial Exhaust Ventilation Deficiency Report Worksheet (Continued)	
Make-up air	**System maintenance**
☐ No replacement air ☐ Insufficient air for dilution of fugitive emissions ☐ Contaminated by exhaust air ☐ Supply diffuser blows on hood face ☐ Supply diffuser blocked ☐ Temperature inadequate ☐ Employee complaints (noise, draft) ☐ Does not meet ASHRAE 62 provisions ☐ Supply not balanced with exhaust	☐ Inadequate maintenance (equipment broken) ☐ Lack of ongoing PM program
Worksite	**Manifold exhaust systems**
☐ Cluttered, housekeeping poor, dirty ☐ Hood positioned near door, window, walkway, other turbulence ☐ Fire escape routes blocked ☐ Aisles blocked	☐ Likelihood of fire/explosion; mixed chemicals ☐ Corrosion in manifold ☐ Condensation in manifold ☐ One hood goes positive ☐ Part of system under positive pressure
Notes	

SMACNA: Sheet Metal and Air Conditioning Contractors' National Association

4.2 Evaluating Sources of Emissions and Exposures to Nanomaterials

4.2.1 Direct-reading Monitoring

Currently, it is unclear which metrics associated with exposures to engineered nanomaterials are most important from a health and safety perspective. The mass-based metric is traditionally used to characterize toxicological effects of exposure to air contaminants. Animal in vivo exposure studies and cell-culture-based in vitro experiments show that size and shape are the two major factors influencing toxicological effects of engineered nanomaterials. Some of the instruments developed to characterize nanoparticles are capable of real-time measurement [Brouwer et al. 2004; Pui 1996; Ramachandran 2005]. Real-time measurement of aerosolized particles, including primary nanoparticles and agglomerates, play an important role in identifying nanomaterial emissions and evaluating control systems during field surveys. The measuring devices used to evaluate controls in the workplace should be portable and robust. Information about readily available instruments and techniques for nanoparticle monitoring (Table 3) has been summarized and discussed in technical reports [BSI 2007a; EU-OSHA 2009; HSE 2004; ISO 2007, 2008; Mark 2007; Park et al. 2010a, b, 2011].

It is noted that some of the instruments on the list in Table 3 are not suited for monitoring nanomaterial emissions in the workplace. For instance, the tapered element oscillating microbalance (TEOM) is used by the Environmental Protection Agency as a standard reference equivalent method to monitor environmental air quality, but the cut-off particle sizes of 10, 2.5, or 1 µm and dimensions of this instrument limit its use for workplace sampling. Another example is the scanning monitoring particle sizer (SMPS), which uses a radioactive source to bring the sampling aerosol to charge equilibrium. This can make shipping difficult. Sometimes it can be difficult to obtain quantifiable mass concentrations of nanomaterials in the workplace using impactor sampling. Newly developed devices, such as photometers, can detect nanoparticles as small as 50–100 nm with resolution around 1 µg/m^3. These instruments can provide continuous monitoring for real-time mass concentrations.

Data from direct-reading instruments only provide a semiquantitative indication of potential nanoparticle emissions. Fluctuating background concentrations may make determination of control efficiency difficult; changes in background concentration may lead the evaluator to think that the controls are performing either better or worse than they are actually performing. In addition, direct-reading instruments cannot distinguish particle source and composition; these can only be determined through off-line microscopic and chemical analysis.

Sampling quality is always an issue for field evaluation. High-quality sampling results can be obtained by following certain steps. The sampling data can only be trusted by using instruments that have been calibrated for nanoparticle sampling before use. Factory calibration for particle counters and sizers typically uses reference materials having a range of particle sizes. If possible, the instruments should be calibrated with the target nanomaterials in the laboratory before using them for field study. The comparison calibration should also be done on identical instruments if they will be used in a field survey. To maintain consistent sampling performance, a zero check for instruments should be performed before daily use and after sampling high-particle emissions. Sampling loss due to particles deposited in sampling tubes can be lowered by using conductive tubing and minimizing tubing length and bends in the tubing. The sampling location should be considered carefully, because nanoparticles diffuse rapidly through the workplace air. The choice of sampling location could have a large influence on the sampling results. The sampling ports must be kept as close as possible to the emission source.

4.2.2 Off-line Analysis

In addition to direct-reading instrument measurements, nanoparticle emissions can also be characterized using off-line analysis techniques. Off-line analysis methods can determine the physical and chemical properties of airborne nanomaterials, such as particle size, shape, surface area, composition, and agglomeration state. These properties are useful to evaluate exposure and toxicology of nanomaterials in the workplace. Off-line analysis can also be useful in separating background nanomaterials from engineered nanomaterials, based on size, shape, morphology, etc.

NIOSH has developed techniques for off-line analysis using filter samples. NIOSH Method 7402 (Asbestos by TEM) was developed to collect filter samples of materials with large aspect ratios for analysis using transmission electron microscopy (TEM) and can be used to determine particle morphology and geometry. NIOSH Method 5040 (Diesel particulate matter as elemental carbon) can be used to measure elemental carbon (e.g., CNT, CNF). Other nanomaterials (e.g., metals) can be collected on filters and analyzed using NIOSH Method 7300 (Elements by ICP). Using the mass determined by chemical analysis and dividing by the total air flow volume will provide a mass concentration of the nanomaterial of interest. As with real-time instrumentation, background samples are collected to help distinguish nanomaterials from incidental ultrafine aerosols. The optical diameters of single particles and agglomerates can be compared to data from direct-reading instruments discussed above.

Filters overloaded with particles cannot be analyzed by direct-transfer TEM analysis. Therefore, filter sample volume needs to be balanced against the particle emission rate to avoid filter overload. The results of the initial walk-through survey with portable particle counters should provide basic information to help determine appropriate filter sampling volume and collection time.

4.2.3 Video Exposure Monitoring

Video exposure monitoring (VEM) is an exposure assessment technique in which real-time monitoring devices (e.g., nanoparticle and dust monitors) are synchronized with video of the work activity [Beurskens-Comuth et al. 2011]. The product of VEM is a video of the work activity with a graphical presentation of exposure concentrations that corresponds to the job task displayed on the video. VEM aides in the identification of work practices that can contribute significantly to overall exposure patterns by giving a visual display of work activities and the corresponding real-time monitoring values. With this exposure assessment tool, both management and employees can be shown which activities have the highest exposure concentrations and can therefore benefit from a change in work practice, installation of engineering controls to mitigate the exposure, or the use of PPE.

The VEM method was initially developed by NIOSH engineers in the late 1980s to bring together work activity data (video recordings) with direct-reading exposure data. By identifying the critical activities that contribute most to a worker's exposure, sampling resources can be directed to controlling those job activities that affect exposures. Work activity variables can also be keyed into the exposure database to statistically assess the impact of work activities on exposures. The method permits researchers and safety and health professionals to capitalize on the time element of the direct-reading data by uniting the exposure measurement with the corresponding work activities. The VEM method allows direct-reading monitors to be used as more than simple detectors and have a significant impact on occupational exposures.

4.3 Evaluating Ventilation Control Systems

Several methodologies are available to evaluate local exhaust systems and other exposure controls. These techniques include indirect approaches, such as the measurement of capture velocity, slot velocities, hood static pressure, and other system performance parameters [Goodfellow and Tahti 2001]. Often these measures are compared with design guidance or standards from organizations such as ASHRAE, ANSI, AIHA, and ACGIH. In general, these tests provide a method of checking system performance without the requirement for expensive instrumentation or a high level of operator experience.

Because these measures do not directly assess system performance, it is often a good idea to use methods that are more specialized than these indirect methods. One method commonly used to evaluate the capture efficiency of the LEV system is the quantitative capture test. Tracer gas release and measurement is a method used to quantitatively estimate the efficiency of industrial exhaust ventilation hoods [Hampl 1984; Hampl et al. 1986; Marzal et al. 2003b]. This method typically involves using a surrogate for the process-generated contaminant and requires the use of special measurement and dispersion equipment to conduct the test. A variety of tracers have been used, including oil mist aerosols, polystyrene latex spheres, and gases [Beamer et al. 2004; Ellenbecker et al. 1983; Hampl 1984].

In addition to the quantitative capture method, qualitative methods, such as smoke release or dry ice tests, are often used to evaluate air movement. Smoke generation and capture is a method often used to qualitatively evaluate the performance of ventilation controls [Marzal et al. 2003a; Woods and Mckarns 1995]. With this method, a source is used to introduce smoke in and around the hood. This allows the researcher to better understand the performance of the hood and evaluate the effect of cross currents on the capture of contaminants. These tests not only give the experimenter a sense of the system performance but provide invaluable information on where other measurements, such as air velocities and tracer gas experiments, should be concentrated. This testing is often conducted while workers are not in the production area, either after the work shift or while workers are on break.

4.3.1 Standard Containment Test Methods for Ventilated Enclosures

Some standard test methods (Table 5) to evaluate fume hoods have been developed: Invent-UK method, DIN 12924, BS 7258, EN 14175:2003, and ANSI/ASHRAE 110-1995. One major difference between ANSI/ASHRAE 110-1995 and other standard test methods is that only one sampling probe is used to detect the test gas concentration near the worker's breathing zone. Other test methods adopt multiple sampling probes connected to a manifold to obtain the area concentration near the fume hood opening. The test methods of DIN 12924 and ANSI/ASHRAE 110-1995 use a manikin to test the containment effectiveness of fume hoods. Dynamic test conditions are specified in the test methods of EN 14175:2003 and ANSI/ASHRAE 110-1995. The purpose of the dynamic test is to evaluate the hood during typical maneuvers such as raising or lowering the sash and simulating the airflow disturbance related to a person walking in front of the hood.

During field evaluations, ventilated enclosures should also be tested during normal-use conditions. Collecting samples both inside and outside the containment opening and in the worker's breathing zone is recommended to assess control effectiveness when workers are performing standard tasks.

Table 5. Comparison of the fume hood performance test methods

Test method	Invent-UK	DIN 12924	BS 7258	EN 14175:2003	ANSI/ASHRAE 110-1995
Country	United Kingdom	Germany	Great Britain	European Union	United States
Test parameters	Face velocity	Tracer gas test	Tracer gas test	Face velocity	Face velocity and cross draft
	Tracer gas test			Tracer gas test	Smoke visualization
				Robustness test (dynamic test by walk-bys and traffic)	Tracer gas test
Tracer gas	10% SF_6 *+ 90% N_2 @ 3.0 LPM	10% SF_6 + 90% N_2 @ 3.33 LPM	10% SF_6 + 90% N_2 @ 2.0 LPM	10% SF_6 + 90% N_2 @ 2–4 LPM	100% SF_6 @ 4.0 LPM
Tracer gas sampling probes	9	20	Multi-probes depending on opening size	Multi-probes (inner and outer grids)	1 in breathing zone
Use of manikin	No	Yes	No	No	Yes

* Sulfur hexafluoride

CHAPTER 5
Conclusions and Recommendations

Engineered nanomaterials are materials that are intentionally produced and have at least one primary dimension less than 100 nanometers (nm). Nanomaterials have properties different from those of the bulk material, making them unique and desirable for specific processes. These same properties may also cause adverse health effects in workers. Currently, the toxicity of many nanomaterials is unknown, but initial research indicates that there may be health concerns related to occupational exposures. Due to the potential for health effects, it is important to control worker exposures to the extent possible. The following are conclusions and recommendations for reducing the potential for employee exposures during nanomanufacturing processes based on current knowledge.

5.1 General

- Hazards involved in processing and manufacturing nanomaterials should be managed as part of a comprehensive occupational safety and health management plan. Preliminary hazard assessments (PHAs) should be conducted to determine the need for control measures during the planning stage. Hazard assessments should be done during the operation of a facility and regularly updated when any processes change.

- The concept of Prevention through Design (PtD) is to design out or minimize hazards early in the design process. When PtD is implemented, the control hierarchy is applied by designing safety into the work environment to prevent work-related injuries and illnesses.

5.2 Control Banding

- With the absence of OELs, control banding is a potentially useful concept in the risk management of nanomaterials. Control banding is not intended to be a substitute for OELs and does not alleviate the need for environmental monitoring or industrial hygiene expertise.

5.3 Hierarchy of Controls

- The hierarchy of controls should be followed when controlling potential occupational hazards from nanoparticles. Elimination and substitution are at the top of the hierarchy. However, eliminating nanomaterials may not be possible as the nanomaterials were likely chosen because of their unique properties. The manner in which these materials are handled and processed can largely affect the overall safety of the process.

- The substitution of less hazardous materials for those that are a higher hazard should be considered to reduce the risk to workers. Substitution also applies to the form of the product used; for example, a slurry with less exposure potential could be used to replace a dry powder.

5.4 Engineering Controls

- If elimination and substitution are not feasible to reduce hazards, engineering controls should be implemented. These could include local exhaust ventilation, isolation measures, and application of water or other material for dust suppression.

- Engineering controls are likely the most effective control strategy for nanomaterials. Common controls used in the nanotechnology industry include fume hoods, biological safety cabinets, glove box isolators, glove bags, bag dump stations, and directional laminar flow booths. Each of these controls should be carefully designed and operated properly to be effective.

- Preventative maintenance schedules should be developed to ensure that engineering controls are operating at design conditions.

- Non-ventilation engineering controls cover a range of controls (e.g., guards and barricades, material treatment, or additives). These controls should be used in conjunction with ventilation measures to provide an enhanced level of protection for workers. Many devices developed for the pharmaceutical industry, including isolation containment systems, may be suitable for the nanotechnology industry.

    a. The continuous liner system allows filling product containers while enclosing the material in a polypropylene bag. This system should be considered for off-loading materials when the powders are to be packed into drums.

    b. Water sprays may reduce respirable dust concentrations generated from processes such as machining (e.g., cutting, grinding). Machines and tooling, as well as the material being cut or formed, must be compatible with water. If a fluid other than water is used, attention should be given to the fluid being applied to avoid creating a health hazard to workers.

- A variety of controls are currently commercially available for use.

- A checklist that collects basic process information (e.g., capacity, location, and usage) and control operation and maintenance parameters can optimize and improve existing exposure control. An example checklist is provided in Table 4.

5.5 Administrative Controls

Administrative controls and PPE are frequently used with existing processes where hazards cannot be effectively controlled solely with engineering controls. This could occur when control measures are not feasible or do not reduce exposures to an acceptable level. Administrative controls and PPE programs may be less expensive to establish but, over the long term, can be very costly to sustain. These methods for protecting workers have proven to be less effective than other measures and require significant efforts by the affected workers. A program that addresses the hazards present, employee training, and PPE selection, use, and maintenance should be in place when PPE is used.

Administrative controls and PPE can also be useful for redundancy, especially in high-hazard situations. While engineering controls serve as primary controls, the administrative and PPE controls provide backup.

Employers should implement the following work practices to control worker exposure to nanomaterials:

- Educate workers on the safe handling of engineered nanomaterials to minimize the likelihood of inhalation exposure and skin contact.
- Provide information to workers on the hazardous properties of the nanomaterials being produced or handled with instruction on how to prevent exposure.
- Obtain the material safety data sheets (MSDS) when using nanomaterials from an outside source and review the information with employees who may come in contact with the materials. Given the lack of complete health information of many nanomaterials, the MSDS may not provide adequate guidance and should be assessed by the health and safety office.
- To reduce the potential for release of nanomaterials, consider transferring powdered materials to a slurry, where possible.
- Clean up spills of nanomaterials immediately and in accordance with written procedures. Appropriate PPE should be donned while performing clean-up tasks.
- Provide additional control measures (e.g., a buffer area, decontamination facilities located by the hazard) to ensure that engineered nanomaterials are not transported outside the work area. Place a sticky mat at the exits of production areas to reduce the likelihood of spreading nanomaterials.
- Encourage workers to use hand-washing facilities before eating, smoking, or leaving the worksite.
- Provide facilities for showering and changing clothes to prevent the inadvertent contamination of other areas (including take-home) caused by the transfer of nanomaterials on clothing and skin.
- Prohibit the consumption of food or beverages in work areas where nanomaterials are handled.
- Ensure work areas and equipment, e.g., balance, are cleaned at the end of each work shift, at a minimum, using either a HEPA-filtered vacuum cleaner or wet wiping methods. Dry sweeping or compressed air should not be used to clean work areas. Cleanup should be conducted in a manner that prevents worker contact with wastes. Disposal of all waste material should comply with all applicable federal, state, and local regulations.
- Store nanomaterials, whether suspended in liquids or in a dry particle form, in closed (tightly sealed) containers whenever possible.
- Conduct routine industrial hygiene and medical monitoring to ensure that work practices and engineering controls are effective.

5.6 Personal Protective Equipment

- Because nanoparticles have been found to penetrate the skin, items such as gloves, gauntlets, and laboratory clothing or coats should be worn when working with nanoparticles. Good hygiene practices for wearing the protective equipment should be followed.

- Gloves made of neoprene, nitrile, or other chemical-resistant gloves should be used and changed frequently or whenever they are visibly worn, torn, or contaminated.
- Respiratory protection should be used to reduce worker exposures to acceptable levels in the absence of effective engineering controls, during the installation or maintenance of engineering controls, for short-duration tasks that make engineering controls impractical, and during emergencies.
- Respirators in the workplace should be used as part of a comprehensive respiratory protection program. The program should include written standard operating procedures; workplace monitoring; hazard-based selection; fit-testing and training of the user; procedures for cleaning, disinfection, maintenance, and storage of reusable respirators; respirator inspection and program evaluation; medical qualification of the user; and the use of NIOSH-certified respirators.

References

60 Fed. Reg. 30336 [1995]. National Institute for Occupational Safety and Health: Respiratory protective devices; final rule. (To be codified as 42 CFR Part 84.)

63 Fed. Reg. 1152 [1998]. Occupational Safety and Health Administration: respiratory protection; final rule. (To be codified at 29 CFR 1910 and 1926.)

ACGIH [2010]. Industrial ventilation: a manual of recommended practice for operation and maintenance. Cincinnati, Ohio: American Conference of Governmental Industrial Hygienists.

ACGIH [2013]. Industrial ventilation: a manual of recommended practice for design. Cincinnati, Ohio: American Conference of Governmental Industrial Hygienists.

Ahn K, Woskie S, DiBerardinis L, Ellenbecker M [2008]. A review of published quantitative experimental studies on factors affecting laboratory fume hood performance. J Occup Environ Hyg 5(11):735–753.

ANSI/AIHA [2012]. Occupational health and safety management systems. Fairfax, VA: American Industrial Hygiene Association Publication No. ANSI Z10–2012.

ASHRAE [1995]. Method of testing performance of laboratory fume hoods. Atlanta, GA: American Society of Heating Refrigerating and Air-Conditioning Engineers, Publication No. ANSI/ASHRAE 110-1995.

ASHRAE [2011]. ASHRAE handbook—HVAC applications. Atlanta, GA: American Society of Heating, Refrigerating, and Air-conditioning Engineers.

Bałazy A, Podgórski A, Gradoń L [2004]. Filtration of nanosized aerosol particles in fibrous filters. I–experimental results. J Aerosol Sci 35:967–980.

Balazy A, Toivola M, Reponen T, Podgorski A, Zimmer A, Grinshpun SA [2006]. Manikin-based performance evaluation of N95 filtering-facepiece respirators challenged with nanoparticles. Ann Occup Hyg 50(3):259–269.

Bayer MaterialScience [2010]. Occupational exposure limit (OEL) for Baytubes defined by Bayer MaterialScience. Leverkusen, Germany: Bayer MaterialScience.

Beamer BR, Topmiller JL, Crouch KG [2004]. Development of evaluation procedures for local exhaust ventilation for United States postal service mail-processing equipment. J Occup Environ Hyg 1(7):423–429.

Bello D, Wardle BL, Yamamoto M, Guzman de Villoria R, Garcia EJ, Hart AJ, Ahn K, Ellenbecker MJ, Hallock M [2009]. Exposure to nanoscale particles and fibers during machining of hybrid advanced composites containing carbon nanotubes. J Nanopart Res 11(1):231–249.

Beurskens-Comuth PAWV, Verbist K, Brouwer D [2011]. Video exposure monitoring as part of a strategy to assess exposure to nanoparticles. Ann Occup Hyg 55(8):937–945.

Brock B [2009]. Knowledge brief: containment hierarchy of controls. Tampa, FL: International Society of Pharmaceutical Engineers.

Brouwer D [2010]. Exposure to manufactured nanoparticles in different workplaces. Toxicology 269(2):120–127.

Brouwer D, van Duuren-Stuurman B, Berges M, Jankowska E, Bard D, Mark D [2009]. From workplace air measurement results toward estimates of exposure? Development of a strategy to assess exposure to manufactured nano-objects. J Nanopart Res 11:1867–1881.

Brouwer DK, Gijsbers JHJ, Lurvink MWM [2004]. Personal exposure to ultrafine particles in the workplace: exploring sampling techniques and strategies. Ann Occup Hyg 48(5):439–453.

BSI [2007a]. Nanotechnologies, part 1: good practice guide for specifying manufactured nanomaterials. Reston, VA: British Standards Institution, Publication No. PD 6699-1:2007.

BSI [2007b]. Nanotechnologies, part 2: guide to safe handling and disposal of manufactured nanomaterials. Reston, VA: British Standards Institution Publication No. PD 6699-2:2007.

BSI [2007c]. Occupational health and safety management systems; requirements. Reston, VA: British Standards Institution, Publication No. BS OHSAS 18001:2007.

Cecala AB, Volkwein JC, Daniel JH [1988]. Reducing bag operator's dust exposure in mineral processing plants. Appl Ind Hyg 3(1):23–27.

Cena LG, Peters TM [2011]. Characterization and control of airborne particles emitted during production of epoxy/carbon nanotube nanocomposites. J Occup Environ Hyg 8(2):86–92.

CFR. Code of Federal Regulations. Washington, DC: U.S. Government Printing Office, Office of the Federal Register.

Chou CC, Hsiao HY, Hong QS, Chen CH, Peng YW, Chen HW, Yang PC [2008]. Single-walled carbon nanotubes can induce pulmonary injury in mouse model. Nano Lett 8(2):437–445.

Conti JA, Killpack K, Gerritzen G, Huang L, Mircheva M, Delmas M, Hathorn BH, Appelbaum RP, Holden PA [2008]. Health and safety practices in the nanomaterials workplace: results from an international survey. Environ Sci Technol 42(9):3155–3162.

Dahm MM, Yencken MS, Schubauer-Berigan MK [2011]. Exposure control strategies in the carbonaceous nanomaterial industry. J Occup Environ Med 53(6 Suppl):S68–73.

Davies CN [1977]. Aerosol science. London: Academic Press, p. 468.

Demou E, Peter P, Hellweg S [2008]. Exposure to manufactured nanostructured particles in an industrial pilot plant. Ann Occup Hyg 52(8):695–706.

DHHS [2009]. Biosafety in microbiological and biomedical laboratories (BMBL) 5th Edition. Cincinnati: U.S. Department of Health and Human Services, Public Health Service, Centers for Disease Control and Prevention, National Institutes of Health, Publication No. DHHS (CDC) 21–1112.

DiNardi SR [2003]. The Occupational environment: its evaluation, control, and management. Fairfax, VA: AIHA Press.

Elder A, Gelein R, Silva V, Feikert T, Opanashuk L, Carter J, Potter R, Maynard A, Ito Y, Finkelstein J, Oberdorster G [2006]. Translocation of inhaled ultrafine manganese oxide particles to the central nervous system. Environ Health Perspect 114(8):1172–1178.

Ellenbecker MJ, Gempel RF, Burgess WA [1983]. Capture efficiency of local exhaust ventilation systems. Am Ind Hyg Assoc J 44(10):752–755.

Eninger RM, Honda T, Reponen T, McKay R, Grinshpun SA [2008]. What does respirator certification tell us about filtration of ultrafine particles? J Occup Environ Hyg 5(5):286–295.

Esco Technologies Inc. [2012]. Pharmacon downflow booth [http://escoglobal.com/products/download/1334055030.pdf]. Date accessed: November 11, 2012.

EU-OSHA [2009]. Literature review: workplace exposure to nanoparticles. Bilboa, Spain: European Agency for Safety and Health at Work, p. 89.

Evans DE, Ku BK, Birch ME, Dunn KH [2010]. Aerosol monitoring during carbon nanofiber production: mobile direct-reading sampling. Ann Occup Hyg 54(5):514–531.

Filtration Group Inc. [2012]. HEPA seal bag in/bag out operation and maintenance manual [www.filtrationgroup.com/../HEPA_BagIn_BagOut_Housing_Install_01.pdf]. Date accessed: November 14, 2012.

Floura H, Kremer J [2008]. Performance verification of a downflow booth via surrogate testing. Pharmaceut Eng 28(6):1–9.

Gao P, Jaques PA, Hsiao TC, Shepherd A, Eimer BC, Yang M, Miller A, Gupta B, Shaffer R [2011]. Evaluation of nano- and submicron particle penetration through ten nonwoven fabrics using a wind-driven approach. J Occup Environ Hyg 8(1):13–22.

Genaidy A, Tolaymat T, Sequeira R, Rinder M, Dionysiou D [2009]. Health effects of exposure to carbon nanofibers: systematic review, critical appraisal, meta analysis and research to practice perspectives. Sci Total Environ 407(12):3686–3701.

Goodfellow H, Tahti E [2001]. Industrial ventilation design guidebook. San Diego, CA: Academic Press, p. 1519.

Hampl V [1984]. Evaluation of industrial local exhaust hood efficiency by a tracer gas technique. Am Ind Hyg Assoc J 45(7):485–490.

Hampl V, Niemela R, Shulman S, Bartley DL [1986]. Use of tracer gas technique for industrial exhaust hood efficiency evaluation—where to sample. Am Ind Hyg Assoc J *47*(5):281–287.

Heim M, Mullins BJ, Wild M, Meyer J, Kasper G [2005]. Filtration efficiency of aerosol particles below 20 nanometers. Aerosol Sci Technol *39*:782–789.

Heitbrink WA, McKinnery WN Jr. [1986]. Dust control during bag opening, emptying and disposal. Appl Ind Hyg *1*(2):101–109.

Hinds WC [1999]. Aerosol technology: properties, behavior, and measurement of airborne particles. New York: Wiley, p. 483.

Hirst N, Brocklebank M, Ryder M [2002]. Containment systems: a design guide. Woburn, MA: Gulf Professional Publishing, p. 199.

HSE [2003a]. Control guidance sheet 301: glovebox. In: COSHH essentials: easy steps to control chemicals. London: Health and Safety Executive.

HSE [2003b]. Control guidance sheet G202: laminar flow booth. In: COSHH essentials: easy steps to control chemicals. London: Health and Safety Executive.

HSE [2003c]. Control guidance sheet G206: sack filling. In: COSHH essentials: easy steps to control chemicals. London: Health and Safety Executive.

HSE [2003d]. Control guidance sheet G208: sack emptying. In: COSHH essentials: easy steps to control chemicals. London: Health and Safety Executive.

HSE [2004]. Nanoparticles: an occupational hygiene review. By Aitken RJ, Creely K S, Tran CL. London: Health and Safety Executive, Health & Safety Executive Publication No. RR 274.

Huang RF, Wu YD, Chen HD, Chen CC, Chen CW, Chang CP, Shih TS [2007a]. Development and evaluation of an air-curtain fume cabinet with considerations of its aerodynamics. Ann Occup Hyg *51*(2):189–206.

Huang S-H, Chen C-W, Chang C-P, Lai C-Y, Chen C-C [2007b]. Penetration of 4.5 nm to aerosol particles through fibrous filters. J Aerosol Sci *38*(7):719–727.

IFA [2009]. Criteria for assessment of the effectiveness of protective measures [http://www.dguv.de/ifa/en/fac/nanopartikel/beurteilungsmassstaebe/index.jsp]. Date accessed: October 18, 2012.

ISO [2007]. Workplace atmospheres—ultrafine, nanoparticle and nano-structured aerosols. Inhalation exposure characterization and assessment. Geneva, Switzerland: International Organization for Standardization, Publication No. ISO/TR 27628:2007.

ISO [2008]. Nanotechnologies—Health and safety practices in occupational settings relevant to nanotechnologies. Geneva, Switzerland: International Organization for Standardization, Publication No. ISO/TR 12885:2008.

Johnson DR, Methner MM, Kennedy AJ, Steevens JA [2010]. Potential for occupational exposure to engineered carbon-based nanomaterials in environmental laboratory studies. Environ Health Perspect *118*(1):49–54.

Kim CS, Bao L, Okuyama K, Shimada M, Niinuma H [2006]. Filtration efficiency of a fibrous filter for nanoparticles. J Nanopart Res *8*:215–221.

Kim SC, Harrington MS, Pui DYH [2007]. Experimental study of nanoparticles penetrations through commercial filter media. J Nanopart Res *9*:117–125.

Kletz T [2001]. An engineer's view of human error. New York: Taylor & Francis, p. 296.

Lee JH, Kwon M, Ji JH, Kang CS, Ahn KH, Han JH, Yu IJ [2011]. Exposure assessment of workplaces manufacturing nanosized TiO2 and silver. Inhal Toxicol *23*(4):226–236.

Lee JH, Lee SB, Bae GN, Jeon KS, Yoon JU, Ji JH, Sung JH, Lee BG, Yang JS, Kim HY, Kang CS, Yu IJ [2010]. Exposure assessment of carbon nanotube manufacturing workplaces. Inhal Toxicol *22*(5):369–381.

Lee KW, Liu BYH [1980]. On the minimum efficiency of the most penetrating particle size for fibrous filters. J Air & Waste Manage. Assoc. *30*(4): 377-381.

Lindeløv JS, Wahlberg M [2009]. Spray drying for processing of nanomaterials. J Phys: Conference Series *170*(1).

Macher JM, First MW [1984]. Effects of air flow rate and operator activity on containment of bacterial aerosols in an class II safety cabinet. Appl Environ Microbiol *48*:481–485.

Maidment SC [1998]. Occupational hygiene considerations in the development of a structured approach to select chemical control strategies. Ann Occup Hyg *42*(6):391–400.

Mark D [2007]. Occupational exposure to nanoparticles and nanotubes. In: Hester RE, Harrison RM, eds. Nanotechnology: consequences for human health and the environment. London: RSC Publishing, pp. 50–80.

Marzal F, Gonzalez E, Minana A, Baeza A [2003a]. Methodologies for determining capture efficiencies in surface treatment tanks. Am Ind Hyg Assoc J *64*(5):604–608.

Marzal F, Gonzalez E, Minana A, Baeza A [2003b]. Visualization of airflows in push-pull ventilation systems applied to surface treatment tanks. Am Ind Hyg Assoc J *64*(4):455–460.

Maynard AD [2007]. Nanotechnology: the next big thing, or much ado about nothing? Ann Occup Hyg *51*(1):12.

McKernan JL, Ellenbecker MJ [2007]. Ventilation equations for improved exothermic process control. Ann Occup Hyg *51*(3):269–279.

Methner M [2008]. Engineering case reports: effectiveness of local exhaust ventilation (LEV) in controlling engineered nanomaterial emissions during reactor cleanout operations. J Occup Environ Hyg *5*(6):D63–D69.

Methner M, Hodson L, Dames A, Geraci C [2010]. Nanoparticle emission assessment technique (NEAT) for the identification and measurement of potential inhalation exposure to engineered nanomaterials—part B: results from 12 field studies. J Occup Environ Hyg 7(3):163–176.

Methner MM, Birch ME, Evans DE, Ku BK, Crouch K, Hoover MD [2007]. Case study: identification and characterization of potential sources of worker exposure to carbon nanofibers during polymer composite laboratory operations. J Occup Environ Hyg 4(12):D125–130.

Mukherjee SK, Singh MM, Jayaraman NI [1986]. Design guidelines for improved water spray systems. Min Eng 38(11):1054–1059.

Nanocyl [2009]. Responsible care and nanomaterials case study. Paper presented at the European Responsible Care Conference, Prague, October 21–23.

Naumann BD, Sargent EV, Starkman BS, Fraser WJ, Becker GT, Kirk GD [1996]. Performance-based exposure control limits for pharmaceutical active ingredients. Am Ind Hyg Assoc J 57(1):33–42.

NIOSH [1997]. Control of dust from powder dye handling operations. Cincinnati: U.S. Department of Health and Human Services, Centers for Disease Control and Prevention, National Institute for Occupational Safety and Health, DHHS (NIOSH) Publication No. 97–107.

NIOSH [2004]. NIOSH respirator selection logic. Cincinnati, Ohio: U.S. Department of Health and Human Services, Centers for Disease Control and Prevention, National Institute for Occupational Safety and Health, DHHS (NIOSH) Publication No. (NIOSH) 2005–100.

NIOSH [2009a]. Approach to safe nanotechnology: managing the health and safety concerns associated with engineered nanomaterials. Cincinnati, OH: U.S. Department of Health and Human Services, Centers for Disease Control and Prevention, National Institute for Occupational Safety and Health, DHHS (NIOSH) Publication No. 2009-125.

NIOSH [2009b]. Qualitative risk characterization and management of occupational hazards: control banding (CB)—a literature review and critical analysis. Cincinnati, Ohio: U.S. Department of Health and Human Services, Centers for Disease Control and Prevention, National Institute for Occupational Safety and Health, DHHS (NIOSH) Publication No. 2009-152.

NIOSH [2011]. Current intelligence bulletin 63: occupational exposure to titanium dioxide. Cincinnati, OH: U.S. Department of Health and Human Services, Centers for Disease Control and Prevention, National Institute for Occupational Safety and Health, DHHS (NIOSH) Publication No. 201-160.

NIOSH [2013]. Current intelligence bulletin 65: occupational exposure to carbon nanotubes and nanofibers. Cincinnati, OH: U.S. Department of Health and Human Services, Centers for Disease Control and Prevention, National Institute for Occupational Safety and Health, DHHS (NIOSH) Publication No. 2013-145.

NNI [no date]. Manufacturing at the nanoscale [http://nano.gov/nanotech-101/what/manufacturing]. Date accessed: October 18, 2012.

Oberdörster G, Oberdörster E, Oberdörster J [2005]. Nanotoxicology: an emerging discipline evolving from studies of ultrafine particles. Environ Health Perspect *113*:823–839.

OECD [2009]. No 11: emission assessment for identification of sources and release of airborne manufactured nanomaterials in the workplace: compilation of existing guidance. Paris: Organisation for Economic Co-operation and Development, ENV/JM/MONO (2009)16.

Paik SY, Zalk DM, Swuste P [2008]. Application of a pilot control banding tool for risk level assessment and control of nanoparticle exposures. Ann Occup Hyg *52*(6):419–428.

Park J, Ramachandran G, Raynor P, Eberly L, Olson G [2010a]. Comparing exposure zones by different exposure metrics for nanoparticles using statistical parameters: contrast and precision. Ann Occup Hyg *54*(7):799–812.

Park J, Ramachandran G, Raynor P, Olson G [2010b]. Determination of particle concentration rankings by spatial mapping of particle surface area, number, and mass concentrations in a restaurant and a die casting plant. J Occup Environ Hyg *7*(8):466–476.

Park J, Ramachandran G, Raynor P, Kim S [2011]. Estimation of surface area concentration of workplace incidental nanoparticles based on number and mass concentrations. J Nanopart Res *13*(10):4897–4911.

Peters TM, Elzey S, Johnson R, Park H, Grassian VH, Maher T, O'Shaughnessy P [2009]. Air-borne monitoring to distinguishing engineered nanomaterials from incidental particles for environmental health and safety. J Occup Environ Hyg *6*:73–81.

Poland CA, Duffin R, Kinloch I, Maynard A, Wallace WAH, Seaton A, Stone V, Brown S, MacNee W, Donaldson K [2008]. Carbon nanotubes introduced into the abdominal cavity of mice show asbestos-like pathogenicity in a pilot study. Nat Nanotechnol *3*:423–428.

Pui DYH [1996]. Direct-reading instrument for workplace aerosol measurements: a review. Analyst *121*:1215–1224.

Ramachandran G [2005]. Occupational exposure assessment for air contaminants. Boca Raton, FL: CRC Press, p. 337.

Ramachandran G, Ostraat M, Evans D, Methner M, O'Shaughnessy P, D'Arcy J, Geraci C, Stevenson, Maynard A, Rickabough K [2011]. A strategy for assessing workplace exposures to nanomaterials. J Occup Environ Hyg *8*:673–685.

Rengasamy S, Eimer BC [2011]. Total inward leakage of nanoparticles through filtering facepiece respirators. Ann Occup Hyg *55*(3):253–263.

Rengasamy S, Eimer BC, Shaffer RE [2009]. Comparison of nanoparticle filtration performance of NIOSH-approved and CE-marked particulate filtering facepiece respirators. Ann Occup Hyg *53*(2):117–128.

Rengasamy S, King WP, Eimer BC, Shaffer RE [2008]. Filtration performance of NIOSH-approved N95 and P100 filtering facepiece respirators against 4 to 30 nanometer-size nanoparticles. J Occup Environ Hyg 5(9):556–564.

Rengasamy S, Verbofsky R, King WP, Shaffer RE [2007]. Nanoparticle penetration through NIOSH-approved N95 filtering-facepiece respirators. Int Soc Respir Prot 24(1/2):49–62.

Roco MC [2005]. International perspective on government nanotechnology funding in 2005. J Nanopart Res 7(6):1–8.

Rossi EM, Pylkkanen L, Koivisto AJ, Vippola M, Jensen KA, Miettinen M, Sirola K, Nykasenoja H, Karisola P, Stjernvall T, Vanhala E, Kiilunen M, Pasanen P, Makinen M, Hameri K, Joutsensaari J, Tuomi T, Jokiniemi J, Wolff H, Savolainen K, Matikainen S, Alenius H [2010]. Airway exposure to silica-coated TiO_2 nanoparticles induces pulmonary neutrophilia in mice. Toxicol Sci 113(2):422–433.

Rushton EK, Jiang J, Leonard SS, Eberly S, Castranova V, Biswas P, Elder A, Han X, Gelein R, Finkelstein J, Oberdörster G [2010]: Concept of assessing nanoparticle hazards considering nanoparticle dosemetric and chemical/biological response metrics. J Toxicol Environ Health 73(5-6):445–461.

Schulte P, Geraci C, Zumwalde R, Hoover M, Kuempel E [2008]. Occupational risk management of engineered nanoparticles. J Occup Environ Hyg 5(4):239–249.

Sellers K, Mackay C, Bergeson LL, Clough SR, Hoyt M, Chen J, Henry K, Hamblen J [2009]. Nanotechnology and the Environment. Boca Raton, FL: CRC Press, p. 296.

Shaffer R, Rengasamy S [2009]. Respiratory protection against airborne nanoparticles: a review. J Nanopart Res 11(7):1661–1672.

Shin WG, Mulholland GW, Kim SC, Pui DYH [2008]. Experimental study of filtration efficiency of nanoparticles below 20 nm at elevated temperatures. J Aerosol Sci 39(6):488–499.

Shvedova AA, Kisin ER, Mercer R, Murray AR, Johnson VJ, Potapovich AI, Tyurina YY, Gorelik O, Arepalli S, Schwegler-Berry D, Hubbs AF, Antonini J, Evans DE, Ku BK, Ramsey D, Maynard A, Kagan VE, Castranova V, Baron P [2005]. Unusual inflammatory and fibrogenic pulmonary responses to single-walled carbon nanotubes in mice. Am J Physiol Lung Cell Mol Physiol 289(5):L698–708.

Smandych RS, Thomson M, Goodfellow H [1998]. Dust control for material handling operations: a systematic approach. Am Ind Hyg Assoc J 59(2):139–146.

Smijs TGM, Bouwstra JA [2010]. Focus on skin as a possible port of entry for solid nanoparticles and the toxicological impact. J Biomed Nanotechnol 6(5):469–484.

Takagi A, Hirose A, Nishimura T, Fukumori N, Ogata A, Ohashi N, Kitajima S, Kanno J [2008]. Induction of mesothelioma in p53+/- mouse by intraperitoneal application of multi-wall carbon nanotube. J Toxicol Sci 33(1):105–116.

Thomas K, Aguar P, Kawasaki H, Morris J, Nakanishi J, Savage N [2006]. Research strategies for safety evaluation of nanomaterials, part VIII: international efforts to develop risk-based safety evaluations for nanomaterials. Toxicol Sci *92*(1):23–32.

Tinkle SS, Antonini JM, Rich BA, Roberts JR, Salmen R, DePree K, Adkins EJ [2003]. Skin as a route of exposure and sensitization in chronic beryllium disease. Environ Health Perspect *111*(9):1202–1208.

Tsai SJ, Ada E, Isaacs J, Ellenbecker MJ [2009a]. Airborne nanoparticle exposures associated with the manual handling of nanoalumina in fume hoods. J Nanopart Res *11*(1):147–161.

Tsai SJ, Hoffman M, Hallock MF, Ada E, Kong J, Ellenbecker MJ [2009b]. Characterization and evaluation of nanoparticle release during the synthesis of single-walled and multiwalled carbon nanotubes by chemical vapor deposition. Environ Sci Technol *43*:6017–6023.

Tsai SJ, Huang RF, Ellenbecker MJ [2010]. Airborne nanoparticle exposures while using constant-flow, constant-velocity, and air-curtain-isolated fume hoods. Ann Occup Hyg *54*(1):78–87.

Vorbau M, Hillemann L, Stintz M [2009]. Method for the characterization of the abrasion induced nanoparticle release into air from surface coatings. J Aerosol Sci *40*(3):209–217.

Walker L [2002]. Process containment design for development facility, part I. Pharmaceut Eng *21*(4):72–75.

Wang J, Liu Y, Jiao F, Lao F, Li W, Gu Y, Li Y, Ge C, Zhou G, Li B, Zhao Y, Chai Z, Chen C [2008]. Time-dependent translocation and potential impairment on central nervous system by intranasally instilled TiO(2) nanoparticles. Toxicology *254*(1–2):82–90.

Wang Y, Wei F, Luo G, Yu H, Gu G [2002]. The large-scale production of carbon nanotubes in a nano-agglomerate fluidized-bed reactor. Chem Phys Lett *364*(5–6):568–572.

Warheit DB, Borm PJ, Hennes C, Lademann J [2007]. Testing strategies to establish the safety of nanomaterials: conclusions of an ECETOC workshop. Inhal Toxicol *19*(8):631–643.

Washington State Department of Labor & Industries [no date]. Industrial ventilation guidelines [http://www.lni.wa.gov/Safety/Topics/AtoZ/Ventilation/default.asp]. Date accessed: October 18, 2012.

Wittmaack K [2007]. In search of the most relevant parameter for quantifying lung inflammatory response to nanoparticle exposure: particle number, surface area, or what? Environ Health Perspect *115*:187–194.

Woods JN, Mckarns JS [1995]. Evaluation of capture efficiencies of large push-pull ventilation systems with both visual and tracer techniques. Am Ind Hyg Assoc J *56*(12):1208–1214.

Woskie S [2010]. Workplace practices for engineered nanomaterial manufacturers. WIREs Nanomed Nanobiotechnol *2*(6):685–692.

WWICS [2011]. The project on emerging nanotechnologies: consumer product inventory [http://www.nanotechproject.org/inventories/consumer/updates/]. Date accessed: October 18, 2012.

Yeganeh B, Kull CM, Hull MS, Marr LC [2008]. Characterization of airborne particles during production of carbonaceous nanomaterials. Environ Sci Technol *42*(12):4600–4606.

APPENDIX A
Sources for Risk Management Guidance

ASSE [2009]. Prevention through design: guidelines for addressing occupational risks in design and redesign processes. ASSE TR-Z790.001. Des Plaines, IL: American Society of Safety Engineers.

Center for Chemical Process Safety [1992]. Guidelines for hazard evaluation procedures: with worked samples. 2nd ed. New York: Wiley-American Institute of Chemical Engineers.

NIOSH [2009]. Approaches to safe nanotechnology: managing the health and safety concerns with engineered nanomaterials. Cincinnati, OH: U.S. Department of Health and Human Services, Centers for Disease Control and Prevention, National Institute for Occupational Safety and Health, DHHS (NIOSH) Publication No. 2009–125 [http://www.cdc.gov/niosh/docs/2009-125/pdfs/2009-125.pdf].

NIOSH [2009]. Qualitative risk characterization and management of occupational hazards: control banding (CB)—a literature review and critical analysis. Cincinnati, Ohio: Department of Health and Human Services, Centers for Disease Control and Prevention, National Institute for Occupational Safety and Health, DHHS (NIOSH) Pub No. 2009–152.

Schulte PA, Geraci CL, Zumwalde RD, Hoover MD, Kuempel ED [2008]. Occupational risk management of engineered nanoparticles. J Occup Environ Hyg 5(4):239–249.

Schulte PA, Geraci CL, Zumwalde R, Hoover MD, Castranova VA, Kuempel E, Murashov V, Vainio H, Savolainen K [2008]. Sharpening the focus on occupational safety and health in nanotechnology. Scand J Work Environ Health *34(*6):471–478.

This page left blank intentionally

APPENDIX B
Sources of Guidance for Control Design

The American National Standards Institute (ANSI) and American Industrial Hygiene Association (AIHA)

ANSI/AIHA Z9.2-2007—Fundamentals Governing the Design and Operation of Local Exhaust Systems. This standard establishes minimum requirements for the commissioning, design, specification, construction, and installation of fixed industrial LEV systems used for the reduction and prevention of employee exposure to air contaminants [ANSI/AIHA 2007].

ANSI/AIHA Z9.5-2003—Laboratory Ventilation. This expanded standard, first published in 1992, includes new chapters on performance tests, air cleaning, preventive maintenance, and work practices, as well as five appendices, such as "Selecting Laboratory Stack Designs," and an audit form [ANSI/AIHA 2003].

ANSI/AIHA Z9.7-2007—Recirculation of Air from Industrial Process Exhaust Systems. This standard established minimum criteria for the design and operation of a recirculating industrial process exhaust ventilation system [ANSI/AIHA 2007].

ANSI/AIHA Z9.9-2010—Portable Ventilation Systems. This standard discusses portable ventilation equipment and systems used for the reduction, control, or prevention of exposure to hazardous atmospheres or airborne substances, and for comfort to employees [ANSI/AIHA 2010].

BSR/AIHA Z9.13—Design, Operation, Testing, and Maintenance of Laminar Flow Fume Hoods. This standard applies to laminar flow fume hoods (LFFH) that use filtered supply air and ducted exhaust to protect products inside the hood from external contamination and exhaust hazardous effluents from the building. This standard provides guidelines for design, operation, testing, and maintenance of laminar flow fume hoods [BSR/AIHA 2010].

The American Society of Heating, Refrigerating, and Air-conditioning Engineers (ASHRAE)

2009 ASHRAE Handbook—Fundamentals. This handbook covers basic principles and data used in the HVAC industry. The ASHRAE technical committees that prepare these chapters strive to provide new information, clarify existing information, delete obsolete materials, and reorganize chapters to make the Handbook more understandable and easier to use [ASHRAE 2005].

2007 ASHRAE Handbook—HVAC Applications. This handbook covers a broad range of facilities and topics and is written to help engineers design and use equipment and systems described in other Handbook volumes [ASHRAE 2007].

ANSI/ASHRAE Standard 110-1995—Method of Testing Performance of Laboratory Fume Hoods). The purpose of this standard is to specify a quantitative and qualitative test method for evaluating the containment of a laboratory fume hood.

The American Conference of Governmental Industrial Hygienists

Industrial Ventilation. A Manual of Recommended Practice. This IH standard reference of ventilation system design and evaluation is made up of two volumes. The first is related to the design of ventilation systems including templates for specific operations. The second provides guidance on the operation and maintenance of ventilation systems and includes information on system performance evaluation.

British Health and Safety Executive (HSE)

COSHH Essentials Control Guidance Sheets. The Generic COSHH Essentials model assigns intervention approaches (i.e., control bands) to workplace tasks after the completion of a semi-quantitative risk assessment. A combination of a substance's toxicity and its inhalation exposure potential determine the desired level of control. After completing this assessment online, users are directed to the appropriate fact sheet. To download the Control Guidance Sheets related to Generic COSHH Essentials go to http://oehc.uchc.edu/news/Control_Guidance_Factsheets.pdf and insert the sheet number in the space indicated. An index to these sheets is shown at that Web site.

The International Organization for Standardization (ISO)

ISO 14644-7:2004—Cleanrooms and associated controlled environments—Part 7: Separative devices (clean air hoods, glove boxes, isolators and mini-environments). ISO 14644-7:2004 specifies the minimum requirements for the design, construction, installation, test, and approval of separative devices, in those respects where they differ from cleanrooms as described in ISO 14644-4 and 14644-5.

The International Society for Pharmaceutical Engineering (ISPE)

ISPE is the world's largest not-for-profit association dedicated to educating and advancing pharmaceutical manufacturing professionals and their industry.

ISPE Good Practice Guide: Assessing the Particulate Containment Performance of Pharmaceutical Equipment. This guide provides a standard methodology for use in testing the containment efficiency of solids handling systems used in the pharmaceutical industry under closely defined conditions. It covers the main factors that affect the test results for specific contained solids-handling systems, including material handling, room environment, air quality, ventilation, and operator technique.